Lecture Notes in Physics

Edited by H. Araki, Kyoto, J. Ehlers, München, K. Hepp, Zürich
R. Kippenhahn, München, H. A. Weidenmüller, Heidelberg
and J. Zittartz, Köln

Managing Editor: W. Beiglböck, Heidelberg

224

Supernovae as Distance Indicators

Proceedings of a Workshop
Held at the Harvard-Smithsonian Center for Astrophysics
September 27–28, 1984

Edited by Norbert Bartel

Springer-Verlag
Berlin Heidelberg New York Tokyo

Editor

Norbert Bartel
Center for Astrophysics, Harvard College Observatory
Smithsonian Astrophysical Observatory
60, Garden Street, Cambridge, Massachusetts 02138, USA

ISBN 3-540-15206-7 Springer-Verlag Berlin Heidelberg New York Tokyo
ISBN 0-387-15206-7 Springer-Verlag New York Heidelberg Berlin Tokyo

Printed in Germany

Printing and binding: Beltz Offsetdruck, Hemsbach/Bergstr.
2153/3140-543210

Preface

This volume contains the proceedings of the workshop "Supernovae as Distance Indicators," held at the Harvard-Smithsonian Center for Astrophysics September 27-28, 1984.

The idea to organize a workshop on this subject came in the spring of 1984 when our progress in angularly resolving an expanding radio supernova with VLBI techniques shed light on the exciting prospect of determining extragalactic distances. The purpose of the workshop was to compare this new method of using supernovae as distance indicators with existing optical methods, namely, the method that is based on the use of Type I supernovae as standard candles, the Baade-Wesselink method for Type I and Type II supernovae, and the ^{56}Ni-radioactivity method for Type I supernovae. Further goals of the workshop were to describe and compare the uncertainties inherent to each of these methods and to discuss the potential that improved theoretical models of supernovae, supernova atmospheres, and young supernova remnants may have for reducing these uncertainties of extragalactic distance determinations.

Allan Sandage opened the meeting with a comparative evaluation of traditional methods of determining distances to remote galaxies. The body of the meeting was devoted to a session on the observational results on supernovae and supernova remnants, followed by sessions on how extragalactic distances can be determined by radio and optical methods.

Although we asked the participants to submit for publication questions and answers posed after each talk, we soon noticed that the informal nature of the meeting and the liveliness of the discussion made this impractical. The few contributions we received were not representative. Therefore, we did not include any discussion sections here.

We originally conceived the workshop as a "Neighborhood" Meeting but very soon realized the need to expand. The workshop was attended by about 20 participants from the CfA and by 25 participants from 20 other institutions.

The meeting was sponsored by the Smithsonian Astrophysical Observatory. We are grateful to B. Bonometti, J. Davis, and C. Gwinn for their help during the workshop and to C. Barrett and K. Brown for their support in organizing the workshop and editing the proceedings.

Norbert Bartel, James Moran
Organizing Committee

January 1985

TABLE OF CONTENTS

3. Radio Supernovae and Inference of Extragalactic Distances

4. Optical Supernovae and Inference of Extragalactic Distances

After-Dinner Talk

CURRENT PROBLEMS OF DETERMINING DISTANCES TO GALAXIES

Allan Sandage[1] and G. A. Tammann[2,3]

[1]Mount Wilson and Las Campanas Observatories of the Carnegie Institution of Washington
[2]Astronomisches Institut der Universität Basel
[3]Visiting Associate, Mount Wilson and Las Campanas Observatories

ABSTRACT

Traditional methods of determining distances to remote galaxies are reviewed to show their relative strengths and weaknesses. Methods that use the luminosity function of globular clusters are rejected as being precise at only the $\pm 1^m$ level. Similar rejection is made of H II region angular diameters. Methods of more promise, in inverse order of precision, are (1) the luminosity function of ScI galaxies, (2) the linewidth-absolute magnitude (Tully-Fisher) method, (3) use of brightest red and blue supergiants, and (4) use of Type I supernovae. Application of each favors $H_o \simeq 50$ km s^{-1} Mpc^{-1} for the Hubble constant.

I. INTRODUCTION

Presently it appears that the most promising route to *accurate* distances to galaxies beyond the M101 Group at $D \simeq 7$ Mpc (precise to, say, 10% per galaxy) is by using supernovae as distance indicators. It is the assessment of this possibility that is the central theme of this conference. We will restrict ourselves to Type I SNe and to the standard candle aspect; other speakers will focus on Type II SNe and other aspects.

This introductory lecture was proposed by the organizers to see why one might come to this startling conclusion. We review here part of the past to point critically to the known weaknesses, as well as the strengths, of the traditional methods used before SNe became quite so fashionable. And it is only fair to remind ourselves that Zwicky had always proposed this route through SN but his acerbic (some would even say scabrous) rejection of all other methods prevented his claim to prevail in earlier days. (But we must also be aware of how primitive our knowledge was then of the dispersion $\sigma(M)$ of the luminosity of Type I SNe at maximum light.) And, despite the sanguine hopes that SNe are excellent distance indicators, we do, in fact, believe that there *are* merits of quite considerable power in the older methods, now to be reviewed.

The problem of determining the Hubble constant breaks into two parts. (1) Precise distances to *very* local galaxies are required, using the only well understood indicator of small luminosity dispersion (Cepheid variables). (2) Secondary distance indicators of higher luminosity, mostly of lower precision, are measured in these local calibrators and then used at distances beyond $D \simeq 25$ Mpc ($m-M = 32^m$; the farther the better) beyond the Virgo cluster velocity perturbation so as to obtain the global value of $H_o = v/D$.

In this review we shall not be much concerned with the absolute distance to the *local calibrators*, as this merely sets the zero point of H_o. We are more concerned with the methods via the secondary indicators that bridge the twilight zone between local and remote distances. However, the problem of the local calibrators *is* central for the *numerical* value of H_o. The principal uncertainty at the moment is the zero point of the Cepheid period-luminosity relation where, although there is general modern agreement to within $\pm 0\overset{m}{.}2$ with the second-generation calibration (Sandage and Tammann 1968, 1969) over that of Wilson's (1939), a recent study (Schmidt 1984) suggests revisions of $\sim 0\overset{m}{.}4$, although this has been challenged (Balona and Shobbrook 1984).

In the present lecture, we adopt our old calibration, justified by the recent discussion by Martin, Warren, and Feast (1979) and more recently by Caldwell (1983).

Five general methods exist for distances beyond the M101 Group where Cepheids are too faint to use. We believe that only three of these have validity and therefore promise. We rate them in the inverse order of their power, dispose of the first two in this section, and discuss the last three in the sections that follow.

(1) The method of brightest globular clusters in external galaxies was early proposed (Sandage 1961) and applied by Racine (1968a, b) and Sandage (1968). An *upper limit* of $H_o = 75$ km s^{-1} Mpc^{-1} was found, with a more probable value of 50 km s^{-1} Mpc^{-1} suggested, also arrived at by de Vaucouleurs (1970) who made explicit the variation of M (bright cluster) with M (parent galaxy), also studied by Hodge (1974). A review of the methods using globular clusters has been given by Hanes (1980).

The difficulty is that the cluster luminosity function does not have a sharp cutoff at bright magnitudes; this makes distances using Q(M) *dependent on the total number of clusters in a given galaxy.* This number is not known unless the distance is already known so as to be able to normalize Q(M) to the total population. The method is almost totally degenerate for that reason (cf., Tammann and Sandage 1983).

Attempts to use the *maximum* of Q(M) and hence its turnover may fare better (cf., van den Bergh, Pritchet, and Grillmair 1985), but the assumptions are untested and are indeed *untestable* until H_o is independently known (i.e., the reliable distance to Virgo must be known as found in other ways). We believe that at the moment we do not know enough to certify the clusters as precision indicators.

(2) Use of angular sizes of H II regions was the first method we tried to piece together the distance scale (Sandage and Tammann 1974a, b, c) beyond M101. Although we found that the linear size increases at almost the degenerate rate with the luminosity of the parent galaxy (op. cit., Fig. 9 of 1974a), we attempted to circumvent the degeneracy by the use of van den Bergh's (1960) luminosity classes (op. cit., 1974a, Fig. 10), thought then to be excellent indicators of parent luminosities. When this was found not to be so (Tammann, Yahil, and Sandage 1979), and implicitly acknowledged by van den Bergh (1980), the application of the direct method of H II region diameters was weakened. However, Kennicut's (1979a, b) more precise data, together with his measured total Hα fluxes, did yield $H_o \simeq 60$ km s^{-1} Mpc^{-1}. But due to the difficulty of distance degeneracy, we here reject this method as being without sufficient merit to compete well with the next three routes.

II. TOTAL GALAXY LUMINOSITIES

A general method with two variations that uses total galaxy luminosities, together with special restrictions, has partial promise. The variations are (1) use of Q(M) for the van den Bergh luminosity classes, with special emphasis on galaxies of Type ScI, and (2) the linewidth (maximum rotational velocity) vs. luminosity methods due to Õpik (1922) and to Tully and Fisher (1977). We shall also discuss a third method that uses the Hubble type plus the van den Bergh luminosity class to form an index (Λ, de Vaucouleurs 1979) that is said to measure luminosity, but we suggest at the end of this section that the method is unsuitable.

a) The Luminosity Function for Spiral Galaxies

Does the luminosity function of galaxies have a maximum, as claimed by Hubble (1936a), or does it increase monotonically with no maximum, as claimed by Zwicky (1957)? Figure 1 shows the debate, reproducing a famous diagram by Zwicky (op. cit.).

It turns out that both luminosity functions are correct; Hubble's applies to spirals (plus Magellanic types of Sdm and Sm) whereas Zwicky's includes low-luminosity dwarf ellipticals (the dE types). We know this from current studies in the Virgo cluster where a complete sample of all galaxies within a 6° radius of the cluster center is known to magnitude $B = 18^m$ ($M_B \simeq -14^m$) (cf., Sandage, Binggeli, and Tammann 1985, hereafter SBT, for a review of the Las Campanas-Basel survey of the Virgo cluster). Figure 2 shows the distribution of apparent B magnitudes for Virgo cluster galaxies, separated into types.

Each distribution has a clear maximum and minimum, as suggested by Hubble. The brightest of the spirals are ScI's with $< B_T >= 10\overset{m}{.}78$ and $\sigma = 0\overset{m}{.}88$.

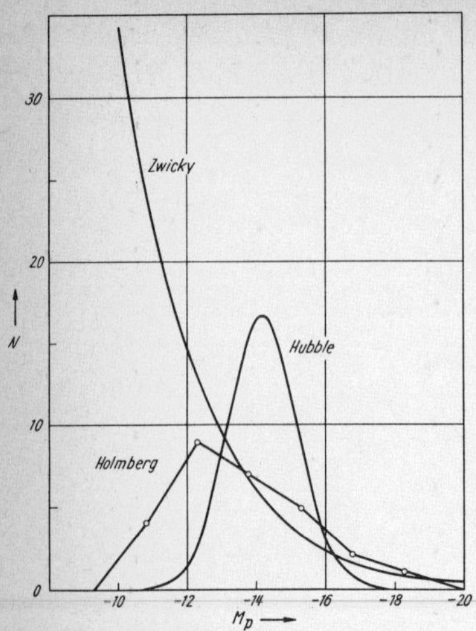

Fig. 1. Comparison of Hubble's luminosity function for galaxies with those of Zwicky and of Holmberg. Our current distance scale is ~ 10 times larger than Hubble's (H_o = 50 km s^{-1} Mpc^{-1} compared with 535 km s^{-1} Mpc^{-1}); hence the modern absolute magnitudes along the abscissa are ~ 5^m brighter.

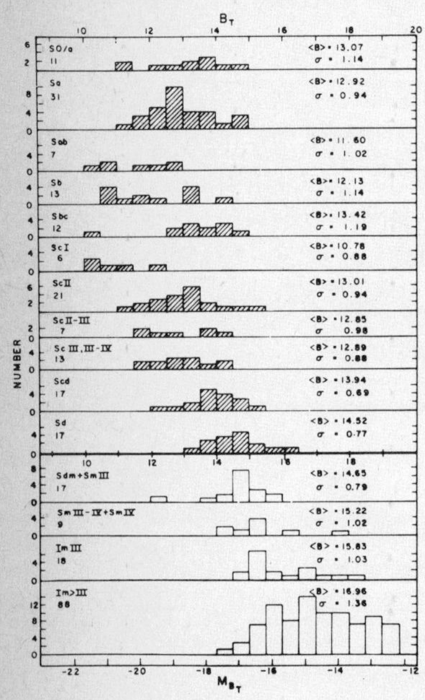

Fig. 2. Distribution of apparent magnitudes for spirals and other star-producing galaxies in the Virgo cluster core from data in the Las Campanas-Basel Virgo cluster survey (Binggeli, Sandage, and Tammann 1985, hereafter BST). Note the near Gaussian character of most of the distributions. The given absolute magnitudes assume a blue apparent modulus of $(m - M)_{AB} = 31\overset{m}{.}7$.

Combining all of the data for spirals gives the distribution shown in Figure 3, which is clearly Gaussian, similar to, but broader than Hubble's.

But if we add the dwarf ellipticals to the spirals in Figure 3 and correct the Virgo catalog for incompleteness, the distribution is that of Figure 4, which is similar to Zwicky's but with a slope not as steep (details are given in SBT).

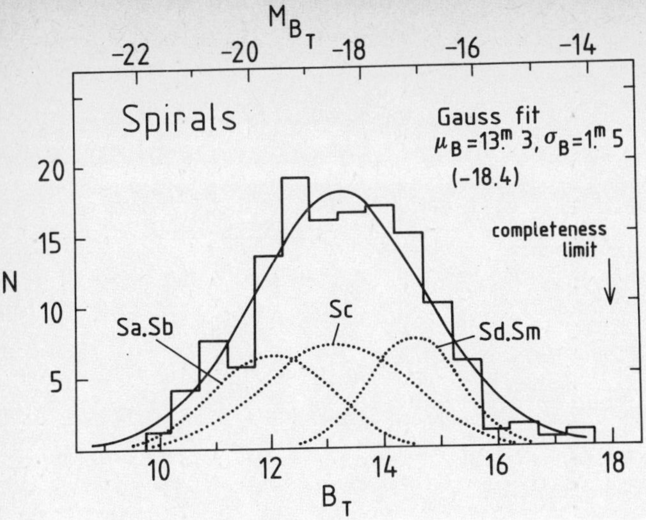

Fig. 3. Combined Q(M) for all spirals in the Virgo cluster catalog. Within the Sc class, narrower distributions apply as a function of luminosity class as shown by Fig. 2.

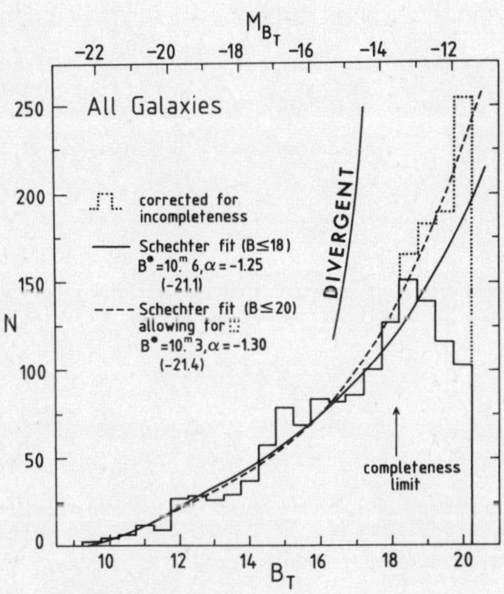

Fig. 4. The luminosity function for all galaxies in the Virgo cluster catalog (BST).

The important point is that Q(M) for ScI galaxies is Gaussian, with moderately small dispersion of *less* than $\sigma(M) = 0^m.88$ in Figure 2 if the faintest of the six examples is neglected. We earlier obtained $\sigma(M) = 0^m.60$ for ScI (Sandage and Tammann 1975, Figs. 1 and 3).

A summary of the luminosity function in the Virgo core, broken into percentage of types, is shown in Figure 5, which shows that there are no dwarf spirals.

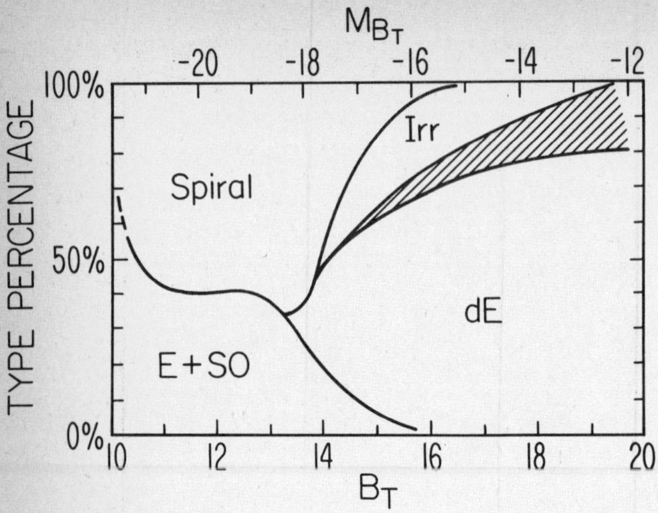

Fig. 5. The percentage of the various morphological types in the Virgo core at different absolute magnitudes. There are no spirals fainter than $M_B \simeq -16^m$. In the faint regime the distribution is dominated by dE and Irr types.

b) The Hubble Constant from ScI Data for Field Galaxies

With, now, a sample of ScI galaxies with its moderately small dispersion in luminosity, we obtain the Hubble diagram of magnitude vs. redshift shown in Figure 6. The slope is 5, as required by a strictly linear velocity-distance relation, and although this slope has been forced through the data by definition, it is a fair approximation. We must expect some Malmquist bias in the sample (Sandage and Tammann 1975, Figs. 2 and 6), but evidently it is small (different catalogs with different magnitude limits were used in the bright and faint regimes).

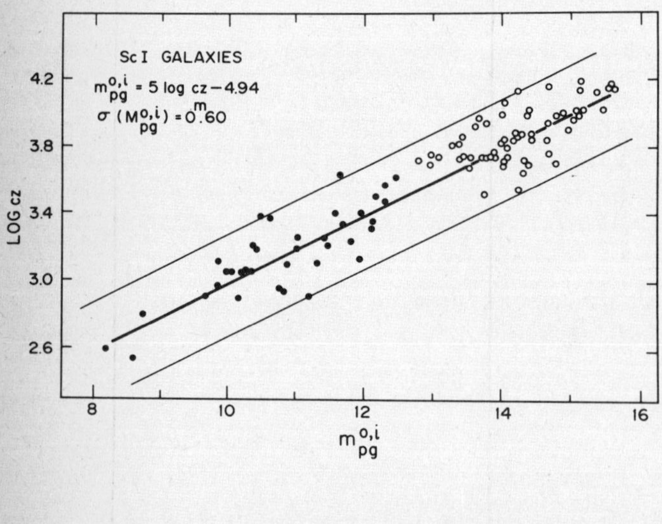

Fig. 6. The sample of ScI galaxies in the general field (Sandage and Tammann 1975) plotted in the Hubble diagram.

The Hubble constant could be obtained from Figure 6 if $< M_B >$ were known for the ridge-line, i.e., if the *mean absolute* magnitude is known for the ScI galaxies in the Figure 6 example. *The weakness of this route to H_o lies in our lack of knowledge of $< M_B >_{ScI}$*. Nevertheless, some limits *can* be placed using the known absolute magnitude of M101 whose distance of $(m-M)_{AB} = 29^m_.2$ is known from Cepheids (Sandage and Tammann 1974b; Sandage 1983) and whose apparent magnitude of $B_T^{o,i} = 7^m_.89$ gives $M_B^{o,i} = -21^m_.31$ for M101 itself. This value, applied to the ridge-line equation shown in Figure 6, using $B - m_{pg} = 0^m_.1$, gives $H_o = 51$ km s^{-1} Mpc^{-1}. The *one sigma* limits on this value, assuming $\sigma(M_B) = 0^m_.60$, are $H_o = 39$ km s^{-1} Mpc^{-1} and $H_o = 67$ km s^{-1} Mpc^{-1}, but clearly we are swimming within the distribution of $M(ScI)$ with no assurance that $M_B = -21^m_.31$ for M101 itself is indeed within 1σ of the ridge-line of Figure 6. Nevertheless, to obtain $H_o = 100$ km s^{-1} Mpc^{-1} requires that M_B for M101 be $2.4\sigma = 1^m_.4$ brighter than the average ScI in Figure 6, and this will occur less than 2% of the time *and is especially not expected of M101*, which is the *nearest* ScI galaxy to us.

c) The Hubble Constant from the ScI Galaxies in Virgo

The five bona fide ScI galaxies in the Virgo core have $< B_T >= 10^m_.45$ with $\sigma(M) = 0^m_.36$. Suppose again that M101, with $M_B^{o,i} = -21^m_.31$, defines the mean value. If so, the modulus of Virgo is $m - M = 10^m_.45 + 21^m_.31 = 31^m_.76$ or $D = 22.5$ Mpc.

The observed mean velocity of $v = 967$ km s^{-1} of Virgo (Kraan-Korteweg 1981) cannot be combined directly with this distance to find H_o because of the deceleration that Virgo exerts on us due to its over-density and also due to the infall of the Virgo supercluster (carrying the Local Group with it) toward the dipole apex of the anisotropy of the microwave background. Correction for these local streaming motions (Tammann and Sandage 1985) shows that the true global value of the Hubble constant via the Virgo cluster itself is

$$H_o = (50 \pm 7)(21.6/D_{VIRGO}) \text{ km s}^{-1} \text{ Mpc}^{-1} \quad ,$$

which, with $D_{VIRGO} = 22.5$ Mpc, gives

$$H_o = (48 \pm 7) \text{ km s}^{-1} \text{ Mpc}^{-1} \quad ,$$

but again the principal uncertainty is equating the luminosity of M101 to $< M >$ for the Virgo ScI galaxies. This uncertainty will always remain until a large enough sample of ScI galaxies is available that has accurate known distances obtained by a precision method, either using Cepheids observed with Space Telescope or, as we shall see in the next sections, using brightest red or blue resolved supergiant stars.

d) The Linewidth-Absolute Magnitude Method

A variation of Öpik's (1922) method of distance determination was discovered by Tully and Fisher (1977) and has been widely used in its various calibrations to obtain distances to galaxies whose 21-cm linewidths are known. The physics of the method is understood at its most elementary level (higher mass means higher rotational velocities and larger linewidths; higher mass usually means higher luminosities), but many subtleties remain unexplained, and although highly promising, the method is not yet mature.

In the hands of different groups, the value of H_o ranges between 100 and 50 km s^{-1} Mpc^{-1}, the high values almost entirely resulting in neglect (or at least in an undercorrection) for the Malmquist bias of any particular sample under study.

To properly correct for statistical effects, one must know the intrinsic dispersion $\sigma(M)$ in absolute magnitude at a given linewidth. Estimates of this quantity vary greatly, from $\sigma(M) \simeq$ $0^m_.3$ to $0^m_.8$.

The relation must also be calibrated, again using the local calibrators that are of different morphological type, and this type-dependence must be taken into account.

As more is learned of the form of the relation as a function of wavelength, many of these problems are expected to be overcome, but at the moment the method is more promising than definitive. For example, when the relation is defined via the *local calibrators* in the blue (B band) and in the infrared (H band at 2μ) and then used to obtain distances to field galaxies, differences in distance moduli exist between the B and H relations that depend on B-H colors, amounting to$\sim 2^m$ over the B-H range (Giraud 1984). Until this is understood, the method cannot be said to be secure.

One final problem to be mentioned is that the local calibrators define the relation near the faint end of the luminosity range for spirals (M_H between -19^m and -21^m), whereas most of the galaxies to which the relation is applied are brighter. Figure 7 shows an application of the IR relation to the Virgo spirals where the difference in the distribution over M_H between the calibration and the unknown is evident. It is clear that the slope must be accurately known to avoid systematic errors in M_H.

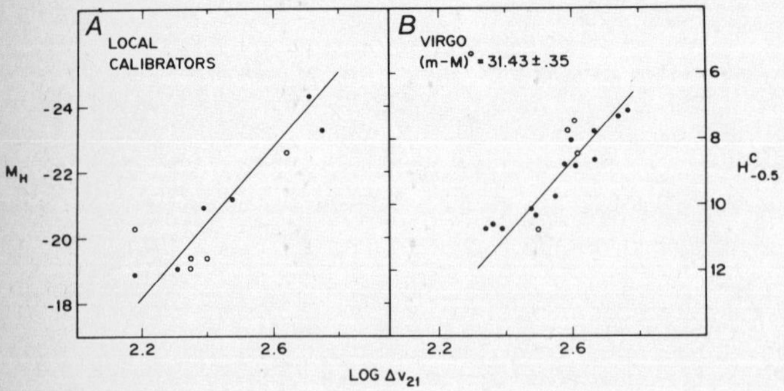

Fig. 7. The Tully-Fisher relation at H magnitudes defined by local calibrators (left) and applied to Virgo cluster spirals to obtain D_{VIRGO} = 19.3 Mpc (Sandage and Tammann 1984).

e) The Hubble Constant from the Tully-Fisher Method

As mentioned, application of the method by a number of workers has given quite inconsistent results. The most recent report of the short-distance-scale ($H \simeq 100$ km s^{-1} Mpc^{-1}) group, who used 21-cm linewidths of ScI galaxies (Bothun *et al.* 1984), quote $H_o = 91 \pm 3$ km s^{-1} Mpc^{-1}. They did apply a correction for Malmquist bias but analysis of the same data by Kraan-Korteweg (1985) shows that their correction is too small. Her result, using also the blue Tully-Fisher calibration, gives $H_o = 58 \pm 4$ km s^{-1} Mpc^{-1}. Such divergent results using the same data and methods show the current uncertainty of the procedure.

A similar conclusion was reached by Sandage and Tammann (1984) using H magnitudes measured by the short-distance-scale group (Aaronson *et al.* 1982; Mould *et al.* 1980) to yield, via Figure 7, a Virgo distance of 19.7 ± 3.1 Mpc and a Hubble constant corrected for all streaming motions (Tammann and Sandage 1985) of $H_o = 55$ km s^{-1} Mpc^{-1}. The only purpose of this demonstration is to show that claims of the superiority of the method over all others (and that $H_o \simeq 100$ km s^{-1} Mpc^{-1} from it) need not be compelling. Indeed, independent data and analysis of groups of galaxies using the blue Tully-Fisher relation has led Richter and Huchtmeier (1984) to a Virgo distance of 24 Mpc, with the resulting global Hubble constant of $H_o = (50 \pm 7)(21.6/D)$ km s^{-1} Mpc^{-1} $= 45 \pm 7$ km s^{-1} Mpc^{-1}.

f) The de Vaucouleurs Λ Index

The short distance scale of de Vaucouleurs (1979) is based on $\Lambda = (T + L)/10$, where T is the quantified Hubble type, and L is the van den Bergh luminosity class.

The distance-limited Virgo cluster complete sample (BST) permits a robust test of the correlation of Λ with apparent magnitude B_T, and hence with absolute magnitude since all galaxies in the Virgo cluster 6° core are statistically at the same distance to within $\delta r/r = 0.1$.

Figure 8 shows the result. There is a correlation, but the dispersion is large at $\sigma(M_B) \simeq 1^m\!.0$ at constant Λ, making the method noncompetitive.

III. BRIGHTEST STARS IN GALAXIES

Beyond the Local Group, Hubble's (1936b) primary distance indicator was brightest resolved stars that he calibrated to have $< M_{pg} > = -6^m\!.1$. At that time this was a brighter absolute magnitude by far than any spectroscopist had assigned to any stellar object.

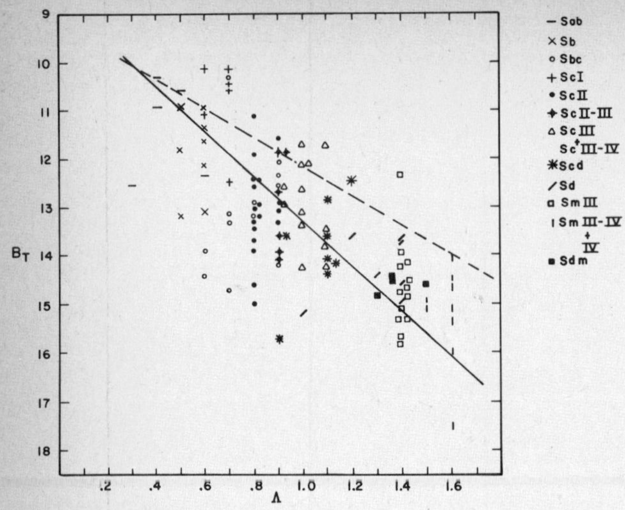

Fig. 8. The de Vaucouleurs (1979) Λ index for spirals in the Virgo cluster 6° core, correlated with apparent magnitude B_T. The dashed line is that adopted by de Vaucouleurs upon which his remote distances are based.

Modern work on the stellar content of nearby galaxies whose distances are known from Cepheids continues to show that the brightest red and blue supergiants have a very small dispersion in absolute luminosity (Sandage and Tammann 1974d; Sandage 1984; Sandage and Carlson 1985). The latest calibrations of the method are shown in Figures 9 and 10.

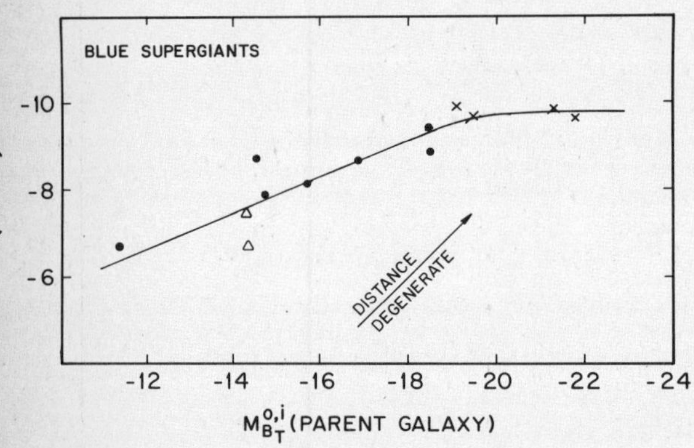

Fig. 9. Absolute magnitude of the three brightest blue supergiants as a function of parent galaxy luminosity. Details are from Sandage and Carlson (1985 with references therein).

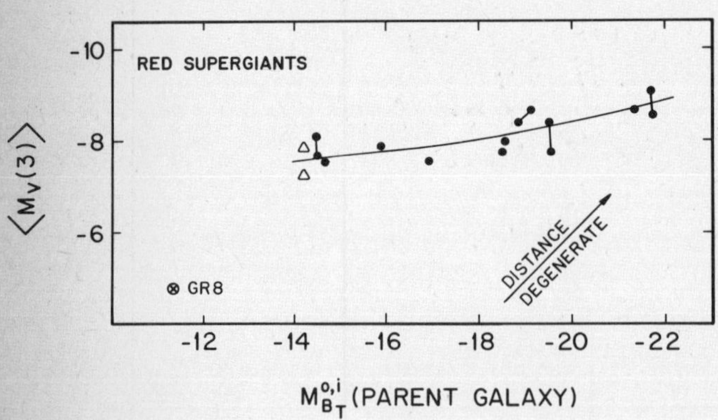

Fig. 10. Same as Fig. 9 but for red supergiants.

The small dispersion about the correlation line, especially for the red supergiants in Figure 10, make these stars at $< M_V(3) > \simeq -8^m$ very promising distance indicators. To use them, we must first identify such stars in external galaxies by showing them to be as red as $B-V \simeq 2\overset{m}{.}0$. They must also be shown to be single. These restrictions limit the method to local distances, closer than $m - M \simeq 30^m$ from the ground, *but* $m - M \simeq 33^m$ *with Space Telescope*, which is just at the near edge of the unperturbed velocity field where H_o must be determined.

The present utility of the brightest star indicator is for the calibration of the luminosity at maximum of Type I supernovae, as explained in a later report here written with R. Cadonau, but its ultimate utility with Space Telescope using the blue supergiants (Fig. 9) should permit the enormous distances of $\simeq$ 150 Mpc to be reached with precision.

IV. TYPE I SUPERNOVAE

That Type I SNe have a very small dispersion in their absolute magnitude at maximum was first demonstrated convincingly by Kowal (1968). A more recent demonstration using only the highest quality photometric data has been given by Cadonau and Tammann (to be published) as used by Sandage and Tammann (1982). The evidence is shown in Figure 11, which is the Hubble diagram for 16 Type I SNe. It is the small dispersion of the points about the line of slope 5 that constitutes the proof. In contrast, present optical evidence suggests that Type II SNe show a considerable spread in their maximum luminosity, making them unsuitable for use as standard candles.

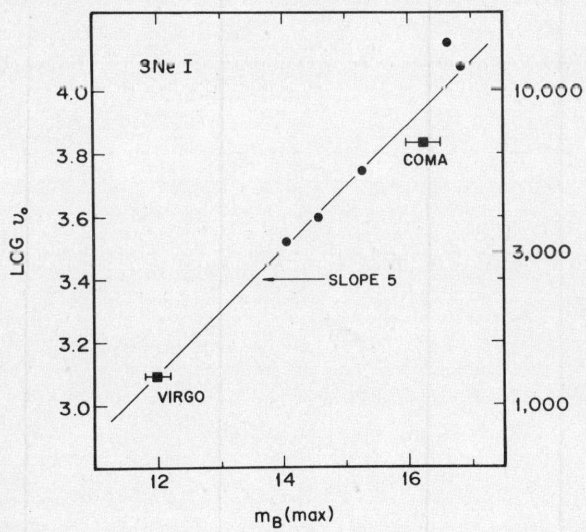

Fig. 11. The Hubble diagram for Type I SNe from data by Cadonau and Tammann.

Once $< M_B(max) >$ is known, the zero-point of the line in Figure 11 leads directly to H_o. As discussed later in this volume, calibration of $< M_B(max) >$ using Figure 10 for the red supergiants in NGC 4214 and IC 4182 led us to a value of $< M_B(max) >= -19\overset{m}{.}74$, which then, via Figure 11, gives

$$H_o = 50 \pm 7 \text{ km s}^{-1} \text{ Mpc}^{-1} \quad .$$

To obtain $H_o = 100$ km s^{-1} Mpc^{-1} would require $< M_B(max) >$ for Type I SNe to be $-18\overset{m}{.}24$, which, at the moment, seems impossible using most external evidence.

V. CONCLUSION

At distances where the Cepheid criteria have been exhausted, only three general methods are known to obtain acceptable distances to more remote galaxies. They are: (1) two variations of the total luminosity of ScI galaxies [Q(M) itself and Tully-Fisher], (2) brightest red and blue supergiants, and (3) Type I supernovae. From the presently available knowledge, Type I SNe promise to be the most precise, as judged from the *internal observational standpoint* alone (Cadonau, Sandage, and Tammann, this volume).

REFERENCES

Aaronson, M. *et al.* 1982, *Ap. J. Suppl.* **50**, 241.

Balona, L. A., and Shobbrook, R. R. 1984, *M. N. R. A. S.* **211**, 375.

Binggeli, B., Sandage, A., and Tammann, G. A. 1985, *A. J.* **90**, in press (BST).

Bothun, G. D., Aaronson, M., Schommer, B., Huchra, J., and Mould, J. 1984, *Ap. J.* **278**, 475.

Caldwell, J. A. R. 1983, *Observatory* **103**, 244.

de Vaucouleurs, G. 1970 *Ap. J.* **159**, 435.

de Vaucouleurs, G. 1979 *Ap. J.* **227**, 380.

Giraud, E. 1984, preprint.

Hanes, D. A. 1980, in *Globular Clusters*, ed. D. Hanes and B. Madore (Cambridge University Press, Cambridge, England), p. 213.

Hodge, P. W. 1974, *Pub. Astr. Soc. Pac.* **86**, 289.

Hubble, E. 1936a, *Ap. J.* **84**, 158.

Hubble, E. 1936b, *Realm of the Nebulae* (Yale University Press).

Kennicut, R. C. 1979a, *Ap. J.* **228**, 696.

Kennicut, R. C. 1979b, *Ap. J.* **228**, 704.

Kowal, C. T. 1968, *A. J.* **73**, 1021.

Kraan-Korteweg, R. C. 1981, *Astr. Ap.* **104**, 280.

Kraan-Korteweg, R. C. 1985, in *The Virgo Cluster*, Proceedings of the September 1984 ESO Virgo Cluster Workshop, ed. B. Binggeli; ESO publication.

Martin, W. L., Warren, P. R., and Feast, M. W. 1979, *M. N. R. A. S.* **188**, 139.

Mould, J., Aaronson, M., and Huchra, J. 1980, *Ap. J.* **238**, 458.

Öpik, E. 1922, *Ap. J.* **55**, 406.

Racine, R. 1968a, *Pub. Astr. Soc. Pac.* **80**, 326.

Racine, R. 1968b, *J. Roy. Astr. Soc. Canada* **62**, 367.

Richter, O.-G., and Huchtmeier, W. K. 1984, *Astr. Ap.* **132**, 253.

Sandage, A. 1961, in *Problems of Extragalactic Research*, ed. G. C. McVittie (Macmillan, New York), p. 359.

Sandage, A. 1968, *Ap. J. Letters*, **152**, L149.

Sandage, A. 1983, *A. J.* **88**, 1569.

Sandage, A. 1984, *A. J.* **89**, 630.

Sandage, A., Binggeli, B., and Tammann, G. A. 1985, in *The Virgo Cluster*, Proceedings of the September 1984 ESO Virgo Cluster Workshop, ed. B. Binggeli; ESO publication (SBT).

Sandage, A., and Carlson, G. 1985, *A. J.* March issue.

Sandage, A., and Tammann, G. A. 1968, *Ap. J.* **151**, 531.

Sandage, A., and Tammann, G. A. 1969, *Ap. J.* **157**, 683.

Sandage, A., and Tammann, G. A. 1974a, *Ap. J.* **190**, 525.

Sandage, A., and Tammann, G. A. 1974b, *Ap. J.* **194**, 223.

Sandage, A., and Tammann, G. A. 1974c, *Ap. J.* **194**, 559.

Sandage, A., and Tammann, G. A. 1974d, *Ap. J.* **191**, 603.

Sandage, A., and Tammann, G. A. 1975, *Ap. J.* **197**, 265.

Sandage, A., and Tammann, G. A. 1982, *Ap. J.* **256**, 339.

Sandage, A., and Tammann, G. A. 1984, *Nature* **307**, 326.

Schmidt, E. G. 1984, *Ap. J.* **285**, 501.

Tammann, G. A., and Sandage, A. 1983, in *Highlights in Astronomy*, ed. R. West (Reidel, Dordrecht), p. 301.

Tammann, G. A., and Sandage, A. 1985, *Ap. J.* , in press.

Tammann, G. A., Yahil, A., and Sandage, A. 1979, *Ap. J.* **234**, 775.

Tully, R. B., and Fisher, J. R. 1977, *Astr. Ap.* **54**, 661.

van den Bergh, S. 1960, *Ap. J.* **131**, 215.

van den Bergh, S. 1980, *Ap. J.* **235**, 1.

van den Bergh, S., Pritchet, C., and Grillmair, C. 1985, preprint.

Wilson, R. E. 1939, *Ap. J.* **89**, 218.

Zwicky, F. 1957, in *Morphological Astronomy* (Springer-Verlag, Berlin), p. 224.

MULTIFREQUENCY OBSERVATIONS OF RECENT SUPERNOVAE

NINO PANAGIA*

Space Telescope Science Institute, Baltimore

1. INTRODUCTION

Supernova (SN) explosions are spectacular events which have called the attention of astronomers since very ancient times: for example, we can find records of SN observations in chinese annals of more than a thousand years ago (Clark and Stephenson, 1977). However, we can place a time limit to modern SN observations around 1885 when the first spectroscopic measurements of S And, or SN 1885a, in the Andromeda galaxy were made. Since then several hundreds of SNe have been discovered in external galaxies and a number of them have been studied in detail both photometrically and spectroscopically. Yet, until a few years ago observations were limited almost exclusively to the optical domain because the existing observing facilities at other wavelengths

TABLE 1
"OLD" FACTS ABOUT SUPERNOVAE
(BASED ON OPTICAL OBSERVATIONS)

	Type I	*Type II*
Where	Elliptical, Spiral and Irregular Galaxies	Only in Spiral Galaxies, associated to the Spiral Arms
How Bright	$M_B(max) \sim -20$ ($L \sim 10^{10}\ L_\odot$)	$M_B(max) < -16$ ($L > 3 \times 10^8\ L_\odot$)
Light Curve	Decay Rate: ~ 0.1 mag/day; $t < 30^d$; ~ 0.01 mag/day; $t > 30^d$	All kinds of behaviour
Optical Spectrum	Broad, unidentified features, more prominent with time $v \cong (10-20) \times 10^3$ km s^{-1}	Balmer Lines HeI, NaI(?), etc. $v \cong (5-10) \times 10^3$ km s^{-1}
Behaviour	Very Homogeneous	Heterogeneous
Progenitors	Low Mass Stars	High Mass Stars

* On assignment from the Astronomy Division, Space Science Department, ESA; on leave from the Istituto di Radioastronomia CNR, Bologna.

(either ground-based such as IR observatories and radio telescopes, or airborne, i.e., satellites, rocket and balloon mounted telescopes) were not sensitive enough for a good detection of the weak fluxes from distant SNe. On the basis of optical information, SNe were subdivided into the two main categories of Type I and Type II supernovae whose properties as on 1978 (e.g. Tammann, 1978) are summarized in Table 1. We see that Type I SNe appear to form a very homogeneous class of objects which are found in all types of galaxies. These characteristics make Type I SNe particularly appealing as standard candles for distance determinations. On the other hand, Type II SNe, which have only been found in spiral galaxies, have properties which vary widely from object to object.

The launch of the International Ultraviolet Explorer (IUE) satellite in early 1978 has marked the beginning of a new era for SN studies because of its capability of measuring the ultraviolet emission of objects as faint as $m_B = 15$ or even fainter. Moreover, just around that time other powerful astronomical instruments have become available, such as the Einstein Observatory for X-ray measurements, the VLA for observations at radio wavelengths and a number of new telescopes either dedicated to infrared observations (e.g. UKIRT and IRTF at Mauna Kea) or equipped with new and highly efficient IR instrumentation (e.g. AAT and ESO observatories). As a result, a wealth of new information has become available which, thanks to the coordinated effort of astronomers operating at widely different wavelengths, has provided us with fresh insights as for the properties and the nature of supernovae of both types. These new results will be summarized and discussed in the following sections.

TABLE 2

IUE OBSERVATIONS OF SUPERNOVAE

Supernova	*Galaxy*	*Type*	*Period of Observation*	*Number of Exposures*
1978g	IC 5201	II	30 Nov–11 Dec 1978	2
1979c	NGC 4321/M100	II-L	21 Apr–4 Aug 1979	29
1980k	NGC 6946	II-L	30 Oct 80–5 Jan 81	33
1980n	NGC 1316	I	11 Dec 80–16 Jan 81	8
1981b	NGC 4536	I	9–11 March 1981	7
1982b	NGC 2268	I	18 February 1982	3
1983g	NGC 4753	I	8–25 April 1983	6
1983n	NGC 5236/M83	I	4 Jul–18 Aug 1983	26
1984	NGC 1559	II(?)	13 Aug 1984	1

2. NEW FACTS ABOUT SUPERNOVAE

2.1 The IUE Sample

Since most of the new information about SN has been either provided or stimulated by ultraviolet observations it is fair to start by presenting a summary of the IUE observations (Table 2). Until now, nine SNe have been observed with IUE, of which 4 were Type II and 5 Type I. However, only four SNe, namely 1979c, 1980k, 1981b and 1983n, were bright enough to obtain high quality ultraviolet spectra and/or to follow their time evolution. Therefore, in the following our discussion will mostly be concerned with these four SNe while referring to the others just briefly. Since the first, and least controversial, SNe observed with IUE in detail were Type II we begin our review with the class of Type II SNe.

2.2 Type II Supernovae

Although there are four Type II SNe which have been observed with IUE, for both the first and the last one the information is very meagre because they were caught only several weeks after maximum and, therefore, they were rather weak. For this reason, the present discussion is limited to the other two SNe (1979c and 1980k) for which the data are quite comprehensive (Benvenuti *et al.* 1982).

Figure 1 displays the UV spectrum and the fluxes in the U, B, V bands of SN 1980k in NGC 6946 obtained on 9 November 1980, i.e. about 10 days after the optical maximum. The UV flux is higher than the extrapolation of the optical spectrum with a black body curve: a clear excess can be seen for $\lambda < 2000$ Å. Such an excess is found at all epochs although it is less pronounced at early times. The characteristics of a high continuous UV emission with an excess at shortest wavelengths was also found for 1979c in NGC 4321 (Panagia *et al.* 1980). Benvenuti, Dopita and D'Odorico (1984) show that the excess may be explained in terms of two-photon emission arising from the upper layers of the ejecta. Alternatively, Fransson (1984) suggests, that the UV excess may be photospheric radiation which has been Compton-scattered by energetic, thermal, electrons ($T \sim 10^9$ K) at the shock front where the ejecta interact with pre-existing circumstellar material. This model is able to explain both the extra radiation observed at short wavelengths and the high ionization implied by some emission lines (e.g. NIV] 1486 Å, CIV 1550 Å, etc.) observed in the spectrum of SN 1979c (Panagia *et al.* 1980; see Fig. 2): therefore, this explanation seems more plausible.

From a comparison of the line profiles observed in the UV and in the visual with theoretical calculations (Fransson *et al.* 1984) it is concluded that the UV emission lines of highly ionized species are produced in the upper atmosphere, as well as Hα or Mg II λ 2800 Å, although just in the outermost layers where density is lower and the ionizing radiation flux is higher. The line

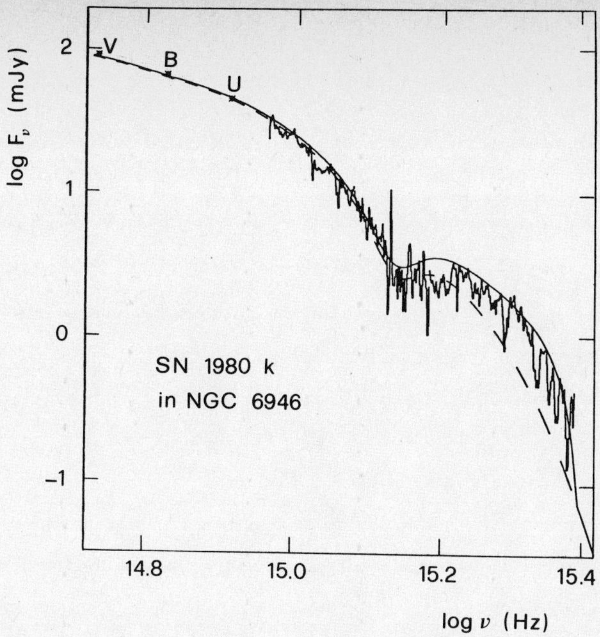

Figure 1. The spectrum of SN 1980k on November 9, 1980. The dashed curve represents the best fit to the optical and near ultraviolet data in terms of a black body at 9900 K and a reddening of E(B–V) = 0.32. The solid line is the spectrum obtained by including also radiation produced by two photon emission (Benvenuti *et al.* 1984).

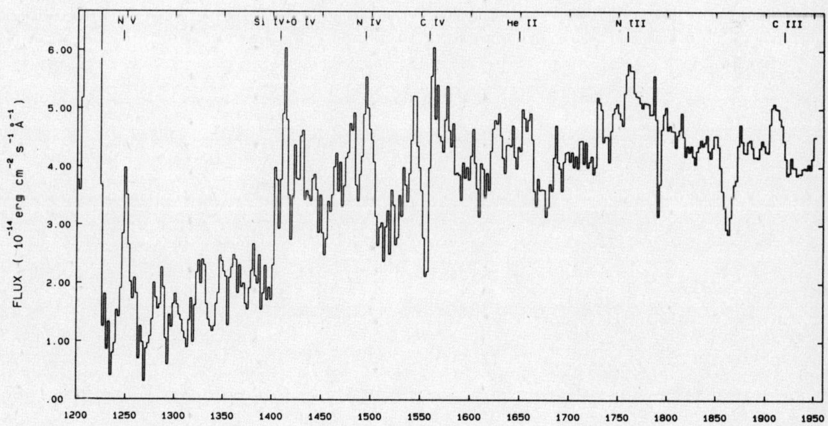

Figure 2. The short wavelength spectrum of SN 1979c on April 24, 1979. The positions of the identified emission lines are shown redshifted by z(NGC 4321) = 0.0054.

profiles imply an expansion velocity of 8400 km s^{-1} (Fransson *et al.* 1984), which is only marginally lower than that measured in the optical (i.e. 9200 km s^{-1}; Panagia *et al.* 1980). From an analysis of the NV 1240 Å, NIV] 1486 Å, CIV 1550 Å, NIII] 1750 Å and C III] 1909 Å line intensities the abundance ratio of nitrogen to carbon has been estimated to be N/C $\sim$ 8 (Panagia, 1980; Fransson *et al.* 1984) i.e. $\sim$ 30 times higher than the cosmic value. This strong enhancement of nitrogen relative to carbon suggests that the pre- Supernova star was a massive supergiant which had undergone a long period of mass loss, thereby exposing CNO processed material.

As for observations at other frequencies, both SNe have been detected in the radio and are currently being monitored with the VLA at wavelengths from 2 to 20 cm (Weiler *et al.* 1981, 1982, 1983, 1984). The time behaviour is qualitatively the same in the two SNe: the "light" curve is characterized by a steep rise, with time scale of a few months and which occurs at earlier times for higher frequencies, followed by a slow decay. The high brightness temperature and the rather steep spectrum at late times ($S_\nu \propto \nu^{-0.7}$) indicate a non-thermal origin for the radio emission. Also, the early and steep rise of the radio emission implies that it originates at the outer layers of the SN envelope because, otherwise, the high f-f- opacity of the ejecta would have prevented any signal from becoming detectable for several years. These data are best explained in terms of a model in which the intrinsic emission is due to the synchrotron process and decreases steadily with time ($\propto t^{-0.7}$). The acceleration of the relativistic electrons may either be produced by turbulence at the shock front of the SN, where the ejecta interact with pre-existing wind material shed by the red Supergiant progenitor (Chevalier, 1981), or be provided by a newly-born plerion (Pacini and Salvati, 1981; Shklovskii, 1981; Salvati, 1983). From a detailed analysis of the radio light curves the former explanation seems to be more likely appropriate. The presence of wind material which is responsible for the f-f absorption accounts for both the steep rising branch of the radio light-curve and the delay of the rise at lower frequencies. From the amount of f-f absorption the mass loss rate of the stellar progenitor is estimated to be about 3×10^{-5} and 4×10^{-6} $M_\odot$ yr^{-1} for SN 1979c and 1980k, respectively, if a velocity of v(RSG) = 10 km s^{-1} is assumed for the stellar wind from the parent red supergiant.

At IR wavelengths SN 1980k is the one Type II SN which has extensively been observed (Dwek *et al.* 1983). During the first couple of months after maximum an IR excess has been revealed for this SN. Such an excess radiation is likely to be the result of extended atmosphere effects in the expanding ejecta. After several months the near IR emission becomes more prominent and is consistent with thermal emission from dust at $\sim$ 800 K. A similar result has been found for SN 1979c at about 260 days past the optical maximum (Merrill, 1980), which also can be interpreted in terms of heated dust (Bode and Evans, 1980; Dwek 1983).

In the X-ray domain, observations with the Copernicus satellite of SN 1979c and with the Einstein Observatory of both SN 1979c and 1980k have been made (Panagia *et al.* 1980; Palumbo *et al.* 1981; Canizares *et al.* 1982). Only the latter has been detected about 40 days after the optical maximum and was already below the detection limit one month later (Canizares *et al.* 1982). The two mechanisms which may explain the observed X-ray emission are: a) Inverse Compton effect on optical photons of the SN, and b) Thermal f-f- emission of the hot shocked gas. Both mechanisms imply a fast decay of the X-ray emission which is consistent with observations. Actually, it is likely that both mechanisms operate to produce X-rays and have comparable importance. The missed detection of SN 1979c at X-ray frequencies (Panagia *et al.* 1980; Palumbo *et al.* 1981) can be explained as due to the quick decline of the X-ray flux which may have gone below the detection limit already at the time of the first Einstein observation (about two months past optical maximum).

Finally, we must mention that SN 1980k was observed with the COS-B satellite from November 4 until December 29, 1980, but no detection was achieved yielding to an upper limit to the flux $F(> 100MeV) < 10^{-6}$ photons cm^{-2} s^{-1} (Cavallo *et al.* 1981). This result indicates that the acceleration of very energetic particles is not efficient at early phases and suggests that a supernova remnant may emit gamma-rays mostly at the end of its life rather than at the beginning (Cavallo *et al.* 1981).

Although qualitatively very similar to each other, these two SNe were considerably different in the absolute luminosity. In fact, after allowance for extinction SN 1979c at maximum appeared to be only about one magnitude fainter than SN 1980k even if the distance to NGC 4321 is almost three times greater than to NGC 6946. Therefore, SN 1979c became more than one magnitude intrinsically brighter than SN 1980k. In general, the spread of properties among different Type II SNe is even larger, not only in absolute luminosity but also in time evolution, i.e. light curve and colours. Therefore, one cannot rely on any of their characteristics to use Type II SNe as standard objects to derive a distance estimate. However, one can take advantage of the fact that hydrogen lines are unambiguously identified and easily measured in the spectrum of Type II SNe to determine the distance by application of the so-called Baade-Wesselink method. In short, it is based on the fact that from spectrophotometric observations one can derive the following parameters: 1) the effective temperature from the shape of the continuum; 2) the apparent flux from direct photometric measurements; 3) the expansion velocity of the photosphere from the Balmer line profiles. Therefore, by determining the first two quantities at a number of epochs one is able to derive the time behaviour of the *angular radius*, which is proportional to the ratio of the photospheric radius to the distance. On the other hand, knowing the expansion velocity one can also estimate the *absolute* value of the radius at any given time and comparing this with the angular radius one can derive a rather accurate estimate of the distance. Since the surface temperature of SNe at early times may be rather high ($> 10^4$ K) a knowledge of the UV continuum is necessary for a reliable determination of the temperature. Combining all available optical and

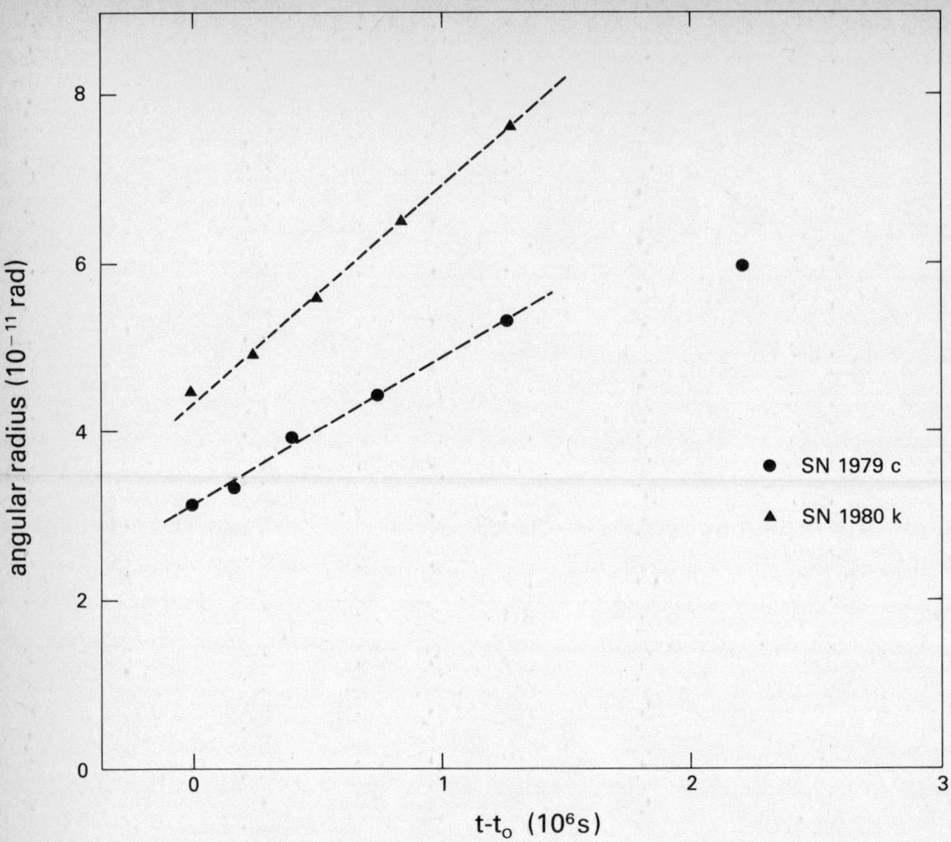

Figure 3. The angular photospheric radii of SN 1979c and SN 1980k are plotted as a function of time. The origin of timc is arbitrary.

UV information, the Baade-Wesselink method has been applied by Benvenuti *et al.* (1984) to the case of both SN 1979c and 1980k assuming the intrinsic spectrum to be a black-body. The derived behaviour of the angular radius as a function of time is displayed in Fig. 3. The angular photospheric radius is found to increase linearly with time at early epochs when the supernova envelope is still very optically thick, but tends to stop rising, and will eventually decrease, when the optical depth will become lower than unity. From the linear part of the curves Benvenuti *et al.* derive angular expansion rates of 1.72×10^{-17} rad s^{-1} and 2.60×10^{-17} rad s^{-1} for SN 1979c and 1980k, respectively, which combined with estimates of the absolute velocity of v_{exp} (1979c) $\simeq$ 9200 km s^{-1} (Panagia *et al.* 1980) and v_{exp} (1980k) $\simeq$ 5700 km s^{-1} give distances of D(1979c) = 17.3 Mpc and D(1980k) = 7.1 Mpc. These values are just intermediate among those one can find in the literature for the two parent galaxies. What is more important, the ratio of the two distances is very close to that determined by other authors even when they estimate significantly different values for the *absolute* distances (e.g. Sandage and Tammann, 1975; de Vaucouleurs, 1979). These results illustrate the power of the Baade-Wesselink method although the problem of its capacity of giving accurate absolute distances still needs to be assessed. Particularly debatable is the problem of computing a set of model atmospheres which enable one to relate the shape of the continuum to the effective temperature and from this to derive an appropriate surface flux.

2.3 "Normal" Type I Supernovae

Until mid-1983 four Type I SNe had been observed with IUE (see Table II). However, none of them was followed for a long time mostly because of pointing constraints of the satellite. Therefore, for all of them we have observations concentrated around the epoch of their maximum light but we know little about their time evolution.

The spectrum is clearly not a smooth continuum but rather displays a number of "bands" which are observed for all of the SNe (cf. Fig. 4). The most prominent feature is the emission which peaks at $\sim$ 2950 Å with a half-power width of $\sim$ 100 Å, i.e. $\Delta v \simeq 10^4$ km s^{-1}. Alternatively, this band may be the result of strong absorptions occuring on both sides of the apparent emission, i.e. centered at $\sim$ 2840 Å and $\sim$ 3060 Å and having half-power widths of the order of 100 Å. A similarly prominent emission band is seen at $\lambda \sim$ 1890 Å in the only one spectrum obtained at short wavelengths for SN 1981b. Several other absorption features can be recognized, which are present at all epochs of observation. Although some of them might be identified with multiplets of Fe I, Fe II and Mg II, no satisfactory identification has been found yet for the majority of the absorptions. Nevertheless, the very fact that the spectrum is virtually the same for the three SNe and at all epochs when observations were made is already an important result. This confirms the homogeneity of properties of Type I SNe which once more show to be all exact replicas of one and the same phenomenon.

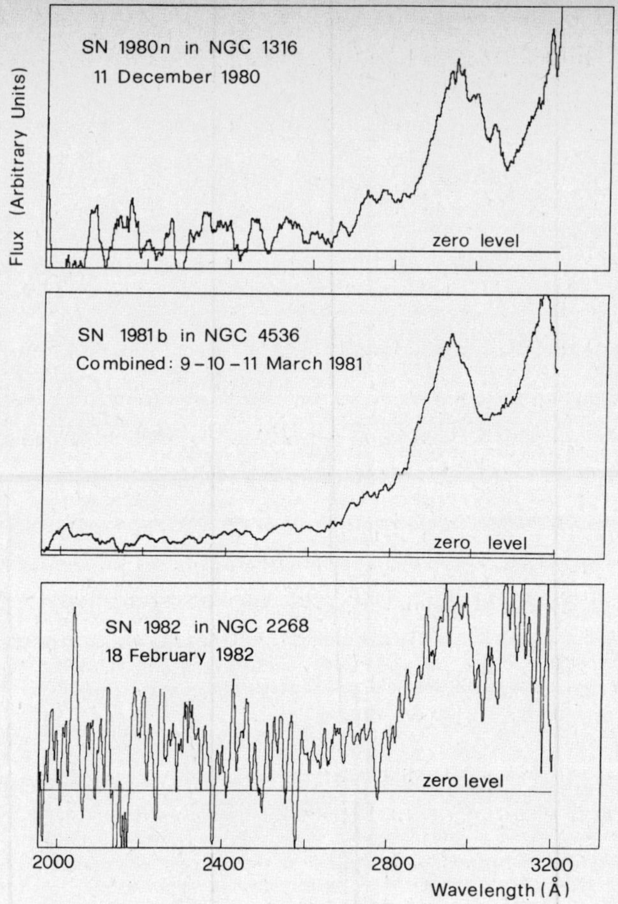

Figure 4. The IUE long wavelength spectra of Type I Supernovae. Note the prominent emission feature centered at ~ 2950 Å.

So far, four Type I Supernovae have been extensively observed in the near IR, i.e. 1972e in NGC 5253, (Kirshner *et al.* 1973); 1980n in NGC 1316: 1981b in NGC 4536 and 1981n in NGC 1316 (cf. Elias *et al.* 1981 and references therein). Two of them, 1980n and 1981b, are in the IUE sample. Again a very close similarity is found in the spectral behaviour and the light variation among the observed SNe. This extends their characteristics of "standard candles" also to the infrared. Another interesting fact is that the light curves in the J, H and K bands display a first minimum 15–20 days after optical maximum, reach a secondary maximum $\sim$ 30 days after maximum and then decay steadily at a rate of $\sim$ 0.04 magnitudes/day. This decay rate is faster than what is observed at optical wavelengths on a long time scale.

Combining different observations, one can reconstruct the complete spectrum of a Type I SN from UV to the IR. As an illustration, the spectrum of SN 1981b at an epoch around the optical maximum is shown in Fig. 5. We see that the ultraviolet spectrum declines very steeply with the wavelength. Moreover, the ultraviolet emission is much lower than a black body extrapolation of the optical spectrum. This is just the opposite of what is found for Type II SNe (cf. Section 2.2). In particular, the UV flux is approximately 10 times lower than the black body curve at the colour temperature appropriate to match the visual spectrum ($T = 15800\ K$, upper curve). Also, in the infrared the observed spectrum is much weaker than a black-body extrapolation from the optical. On the other hand, the IR and UV spectra can smoothly be connected to each other by a lower temperature black-body ($T \sim 9400\ K$, lower curve). These results indicate that the opacity in both UV and IR must be much higher than at optical wavelengths, so that the radiation temperature is close to the effective temperature in the visual but reduces to the minimum temperature $T_{min} \sim 0.6\ T_{eff}$ both at shorter and at longer wavelengths.

Within this sample of Type I SNe only SN 1981b has been observed in the radio. It has been monitored at 5 GHz with the VLA at regular time intervals within the period March 1981–February 1983 but it has never been detected (Weiler *et al.* 1984), the upper limit to the radio flux being as low as 50 μJy (1 σ). From a comparison with the radio properties of Type II SNe, Panagia (1984a) argues that the lack of radio emission from SN 1981b implies that its progenitor was a star which in the late phase of its evolution underwent a low mass loss ($\dot{M} < 10^{-7}\ M_{\odot}\ yr^{-1}$) and, therefore, was a relatively low mass star ($< 3\ M_{\odot}$).

2.4. Supernova 1983n in NGC 5236: An Anomalous Type I

On July 3rd, 1983 a bright supernova ($m_v \sim 13$ at discovery) was discovered in the southern spiral galaxy NGC 5236 = M83. The IUE observations were started on July 4th, i.e. one day after discovery, and continued at regular intervals until mid-August when the SN became too faint to be detected. The first long wavelength IUE spectrum obtained on July 4th was remarkably similar to that displayed by SN 1981b on 9–11 March, 1981, i.e. at the epoch of its maximum in the B band

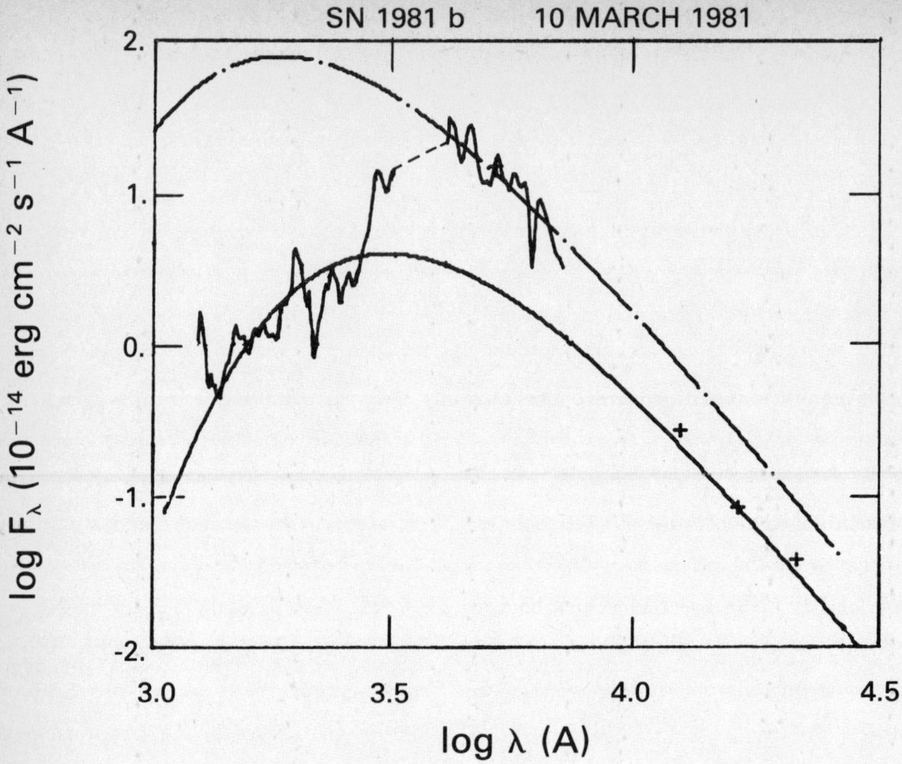

Figure 5. The spectrum on the Type I SN 1981b on March 10, 1981, dereddened with E(B–V) = 0.24. The upper curve is a black body at 15800 K which fits the optical spectrum. The lower curve is a black body at 9400 K which connects the IR and UV spectra.

(Barbon, Ciatti and Rosino, 1982). In particular, the spectrum showed the prominent "emission" feature centered at ~ 2950 Å as well as a sharp decline of the flux with increasing frequency, which were found to characterize the spectra of all of the four Type I SNe observed before (Panagia 1982, 1984b). For this reason and also on the basis of the general appearance of the optical spectrum it was immediately classified as a Type I.

The close similarity with the UV spectrum of SN 1981b was still preserved for the spectrum of July 5th but on the 7th of July a strong "absorption" feature developed which extended from ~ 2550 Å to ~ 2700 Å. It remained present in the spectrum at least until late July when the fading of the intrinsic flux of the SN relative to the background emission due to an angularly nearby OB association made it impossible to reveal reliably any absorption feature in the UV spectrum. The maximum emission at UV wavelengths occurred between the 8th and 12th of July. This is a first hint of the strangeness of this SN in that its UV spectrum at an epoch well prior to maximum closely resembled that of other Type I SNe at the maximum and became markedly different at later epochs. A similarly unusual behaviour has also been observed in the visual where the gross aspect of the SN 1983n spectrum resembled that of Type I SNe but, at a closer inspection, the observed features were not quite the same as in "normal" Type I SNe (e.g. SN 1981b): the 6150 Å absorption was completely missing and the other features vaguely resembled those of much more evolved Type I SNe (Panagia *et al.* 1984). In simple terms, SN 1983n looked like it was "born old" for both UV and optical spectral characteristics. The second "surprise" came from a comparison of optical and infrared data: while the other Type I SNe had an IR emission significantly lower than a black body extrapolation of the optical spectrum (Panagia 1984a), SN 1983n displays an emission which at all epochs can almost perfectly be fitted to a black body curve for all wavelengths longer than, say, 4000 Å (see Fig. 6). In fact, judging from the observed ratios F(2950 Å)/F(FES) and F(J)/F(FES), this supernova is found to be both the reddest and the bluest Type I SN among those ever observed with IUE. This fact, together with direct observations of interstellar absorption lines (Jenkins *et al.* 1984), indicates that the extinction toward SN 1983n is rather modest, i.e. $A_v < 1^m$. This leads to the third surprise: the FES light curve reached a maximum on $\sim 19th$ July at m(FES) $\simeq 11.5$ which, corrected for reddening and adopting the largest distance among those proposed for M83 (~ 8 Mpc; Sandage and Tammann, 1974), corresponds to M(FES) $= -18.5$. This makes it at least 1.3 magnitudes fainter than average Type SNe, which is an enormous deviation for alleged standard candles!

The detailed light curves from UV to IR are displayed in Fig. 7 in the form of λF_λ as a function of time. The IR data were collected by Allen *et al.* at several collaborating observatories. As already apparent in Figure 6, most of the energy at all epochs is emitted at optical wavelengths. In particular, only 13% of the total luminosity was emitted shortward of 3400 Å at the epoch when the UV emission attained a maximum. Moreover, there is no indication of any stronger emission

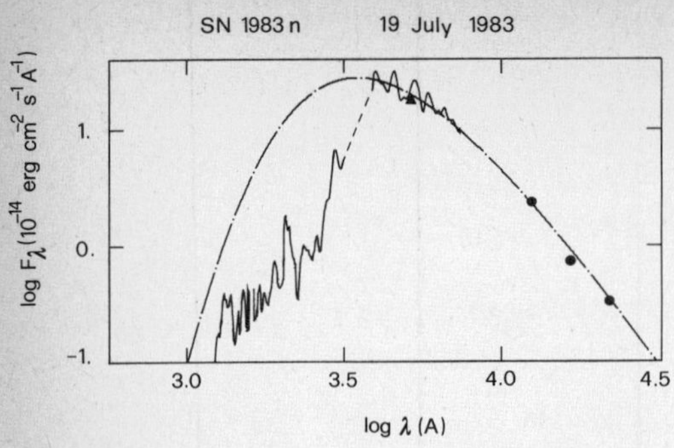

Figure 6. The spectrum of SN 1983n on July 19, 1983, dereddened with E(B–V) = 0.16. Both the UV and the optical spectra have been smoothed with a 100 Å bandwith. The triangle represents the FES photometric point, the dots represent the J, H and K photometric data. The dashed-dotted curve is a black body curve at T = 8300 K.

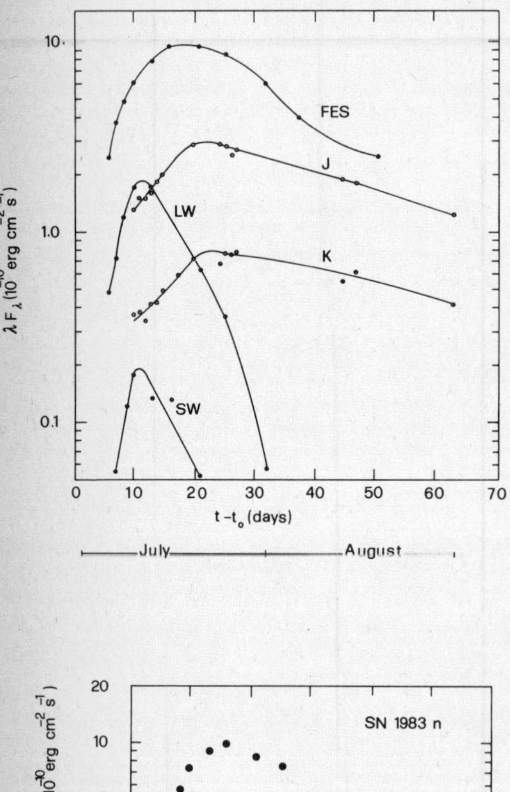

Figure 7. The light curves from the UV to the IR, in the form of log λ F_λ as a function of the time elapsed since the explosion. The curves denoted as SW and LW represent the behaviour of the average flux measured with short and the long wavelength IUE cameras, respectively.

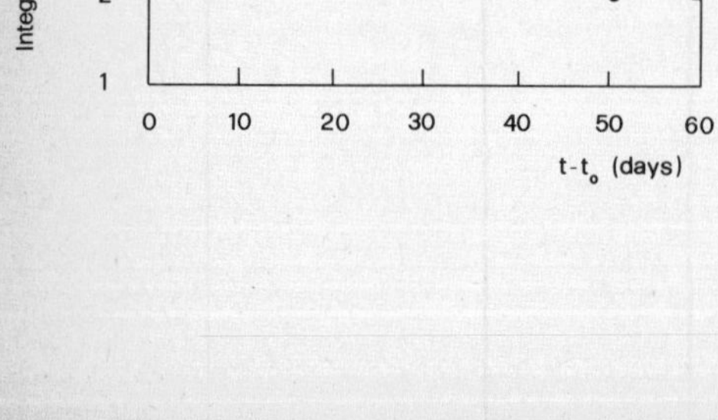

Figure 8. The bolometric light curve obtained by direct integration of the spectra dereddened with E(B–V) = 0.16.

in the UV at very early epochs: this implies that the initial radius of the SN, i.e. the radius the stellar progenitor had when the shock front reached the photosphere, was definitely much less than 10^{13} cm and probably lower than 10^{12} cm, and rules out a red supergiant as a possible progenitor of this supernova.

The bolometric light curve presented in Figure 8 is computed by direct integration of the spectrum over all wavelengths: this is the first time that this can be done reliably for a Type I SN. Both the FES and the bolometric light curve are found to be broader than observed in other Type I SNe and both the initial rising phase and the decline after maximum to be slower than average. From the shape of the light curve the epoch of the explosion is estimated to be the 29th of June with a possible uncertainty of ± 1 day. A comparison of the bolometric light curve with the results of model calculations involving the radioactive decay of ^{56}Ni into ^{56}Co as the main source for the luminosity (e.g. Arnett, 1982) leads to estimates of the ^{56}Ni mass in the range 0.1–0.5 $M_\odot$ for an adopted distance of 8 Mpc (and much less if the distance is smaller). This implies that models which involve the complete incineration of a white dwarf to transform it entirely into ^{56}Ni can be ruled out for this supernova.

SN 1983n was also observed at radio wavelengths with the VLA, starting as early as July 6th and following the event for several months (Sramek, Panagia and Weiler 1984). Contrary to any "sane" expectation, the SN was immediately detected at 6 cm with S(6 July) = 2.0 $\pm$ 0.5 mJy. The maximum flux density was detected on July 27th (S = 28 mJy). Since then, the flux declined steadily with a power law decay of the type $S \propto t^{-1.6}$. Observations made at 20 cm suggest a power law spectrum with a strongly non-thermal spectral index $\alpha \sim 1$. This is the first detection of a Type I SN at radio wavelengths. The success is partly due to the fact that the parent galaxy is relatively nearby but also to an intrinsic radio luminosity exceptionally high for a Type I SN. In fact, its radio emission is at a comparable level as it is observed in Type II SNe (i.e. 1979c and 1980k) with the difference that in SN 1983n the time decay is faster, $t^{-1.6}$ instead of $t^{-0.7}$, and the spectrum is steeper (spectral index 1 instead of ~ 0.7; Weiler *et al.* 1984). In order to both produce relativistic electrons at the shock front and to account for the initial strong absorption, the presence of a circumstellar envelope is needed, whose extent implies a steady mass loss rate of $\sim 3 \times 10^{-6}$ $M_\odot$ yr^{-1} for an assumed wind velocity of 10 km s^{-1} (Chevalier, 1984, Sramek *et al.* 1984). On the other hand, the SN progenitor could not be responsible for this wind because the bolometric light curve excludes that it was a red supergiant. The most plausible picture is that of a SN produced in a binary system: both stars had to be originally rather massive (> 5 $M_\odot$) with the primary, which eventually goes supernova, about 1–2 $M_\odot$ more massive than the companion. The primary evolves first and ends up as a massive white dwarf, a little below the Chandrasekhar limit ($M_{wd} \sim 1.35$ $M_\odot$). It slowly accretes mass from the companion which, meanwhile, has reached the red supergiant phase. Thus, when the WD exceeds the critical mass and explodes as a SN, the explosion occurs within the envelope of the companion: the observed radio emission originates from the interaction of the ejecta with that envelope. Within this frame, it is clear that a strong radio emission from Type I SNe is possible only when the involved binary system is relatively massive and, therefore, as such it must be a rather rare event.

2.5 Anomalous Type I: Recognition of a Subclass

Although rare, "anomalous" Type I SNe like 1983n are not a totally uncommon occurrence in spiral galaxies. For example SN 1962e in NGC 1073 (Bertola, 1964) and SN 1964l in NGC 3938 (Bertola *et al.* 1965) displayed characteristics very similar to those of SN 1983n: the 6150 Å absorption was missing, the light curve evolution was rather slow and the magnitude at maximum was at least 2^m fainter than average. In fact, considering the frequency of Type I SNe as a function of the absolute magnitude (Tammann, 1982; see Fig. 9) one finds that, while in elliptical and S0 galaxies the distribution is sharply peaked around $M_B \sim -20^m$ with a small dispersion, $\sigma \sim 0\overset{m}{.}4$, for spiral galaxies the distribution still has a main peak around $M_B \sim -20^m$ but extends considerably toward low luminosities. This extension is centered around $M_B \sim -18^m$ and contains about 1/3 of the total in a sample of 34 well studied Type I SNe in spiral galaxies. Moreover, these objects are characterized by having considerably redder colours than average, just the same as found for SN 1983n. It has been suggested that the fainter magnitude of these SNe could be due to heavy reddening, thus justifying also the red colours. However, the level of extinction systematically needed to cause an average shift of ~ 2 magnitudes is unreasonably high for such a large fraction of SNe. In addition, we know for sure that in the case of SN 1983n the extinction is modest and that the luminosity is intrinsically low. Therefore, it is more likely that spiral galaxies contain a sub-class of Type I SNe which does not exist in early type galaxies. They would represent the end products of the evolution of rather massive binary systems in which neither of the components was massive enough to explode as a Type II SN.

The large spread in their absolute luminosity makes Type I SNe in late type galaxies unsuited to be used as standard candles. In principle, spectroscopic measurements should be made to clarify if a SN is of either a "normal" or an "anomalous" kind. However, it is not even clear yet whether the Type I SN distribution consists of the combination of two sharply separated groups or rather there is a gradual transition between fully normal to extremely anomolous subtypes, which would make the distribution intrinsically much broader and ill-defined.

3. CONCLUSION

From what we have seen in the previous sections, it is clear that Type II SNe can be used for distance determination only through the Baade-Wesselink method. To derive meaningful distances with this method one needs an extensive time coverage of the event, say from maximum light

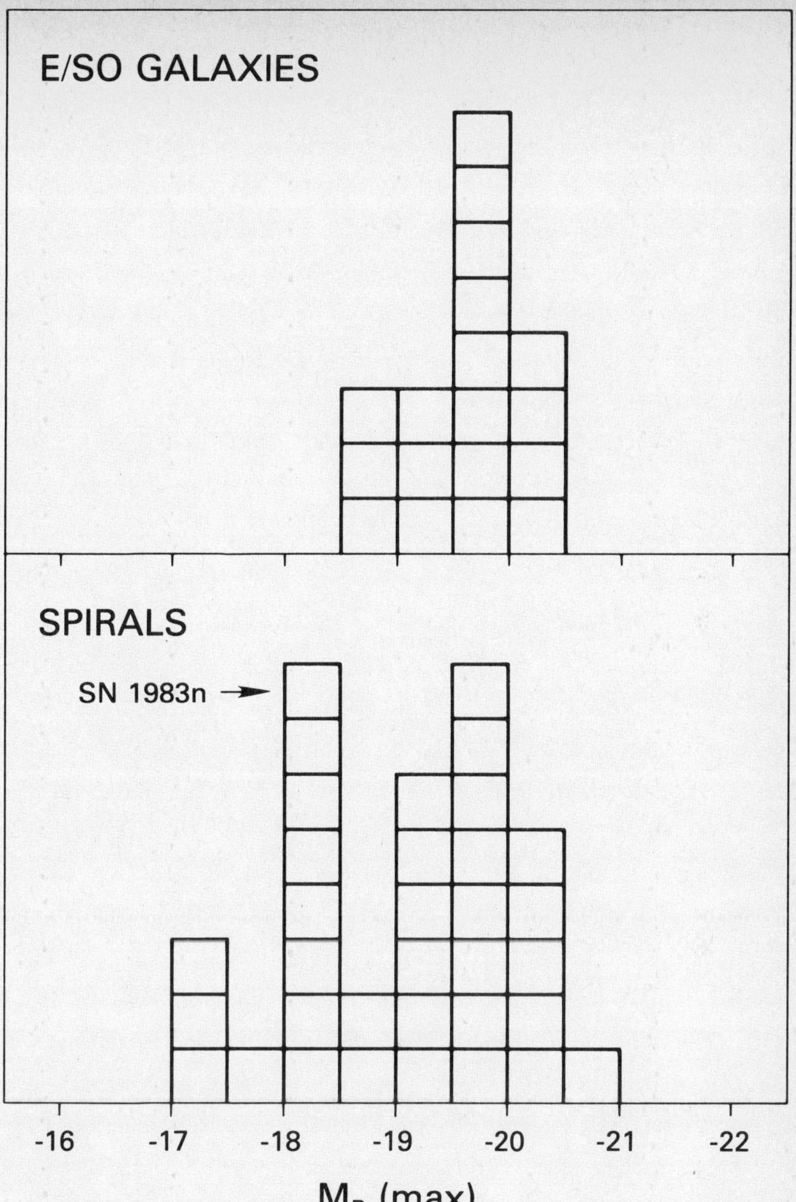

Figure 9. The distribution of Type I SNe as a function of the absolute B magnitude for E/S0 galaxies (upper panel) and for spiral galaxies (lower panel). The box containing SN 1983n is explicitely marked.

until about six weeks—two months after maximum, with both photometric *and* spectroscopic observations from the UV to the IR: these measurements are required to provide information on the integrated flux, the effective temperature and the velocity in the photosphere (see section 2.2). Because of the need of spectroscopic measurements at advanced stages of their evolution (i.e. when the absolute magnitude may have declined to $M_B \sim -15$) one cannot expect to study very far away Type II SNe: in fact, even by using the prism of the Faint Object Spectrograph, mounted on the Space Telescope, which can take spectra of objects as faint as $\sim 26^m$ with a resolution of $\lambda/\Delta\lambda \sim 10^2$, one would be limited to SNe with distance moduli of $\sim 40^m$ which correspond to a redshift limit of $z_{max} \sim 0.2$. It is clear, then, that Type II SNe may be used to determine H_o but can hardly provide any information on q_o.

For different reasons, the situation is not much better with Type I SNe. On the average they are brighter than Type II SNe and, therefore, one may hope to reach farther distances, say by a factor of two or even more. Also, one may think to determine distances with Type I SNe both by using the Baade-Wesselink method or by taking them as standard candles. However, the former method requires a *good* and reliable identification of the observed spectral features and a set of appropriate model atmospheres to interpret the data: for both aspects our knowledge is, to be generous, rather poor and, therefore, this method does not seem to be feasible at present. The use of Type I SNe as standard candles is by far a more promising and straightforward method: it is based on the assumption that both the luminosity of Type I SNe at maximum and the shape of the light curve are the same in all events. All it would require is a series of *photometric* measurements to determine the light curve and from there to infer the magnitude at maximum. To do this one may rely on the sensitivity of the Faint Object Camera and/or the Wide Field/Planetary Camera on ST which both can reach the 28th magnitude in a viable exposure time. The high sensitivity of these instruments combined with the intrinsically brighter luminosity should allow one to cover events with a distance modulus as high as 45^m which corresponds to $z \sim 1$. However, there are a number of difficulties which limit us to closer distances. First of all, one has to ascertain whether the intrinsic properties of Type I SNe vary with z and, if so, exactly how. Secondly, there is the problem of the anomalous Type I SNe which have been shown to pester the Type I population in spiral galaxies. In order to determine the exact subtype of a SN event one needs to make a *series* of spectroscopic measurements: this necessity will limit the observations of SNe to $z \sim 0.1 \div 0.2$. Nor the situation improves much with Type I SNe in elliptical galaxies despite the fact that they are predominantly made of low mass stars and, therefore, they should not incur into the trouble of an anomalous subtype of SNe I. The problem is that when looking at objects far away one also looks back in time. Now, it is possible to show (Panagia 1984c) that in order to be able to exceed the Chandrasekhar limit by accreting matter from the companion, while being in the white dwarf stage, and go supernova, the primary star must have had an *initial* mass definitely greater than a solar mass, say $M_{in} \geq 1.5 - 2\ M_\odot$. Therefore, the frequent occurrence of Type I SNe in elliptical galaxies implies that star formation was still quite active back at about one lifetime of a $\sim 1.5\ M_\odot$ star, i.e. $\sim 2.5 \times 10^9$ years ago. The corresponding redshift for such a look-back time is $z \sim 0.2$ for $H_o = 50$ km s^{-1} Mpc^{-1} and $z \sim 0.5$ for $H_o = 100$ km s^{-1} Mpc^{-1}. Thus, when looking at elliptical galaxies at redshifts larger than a few tenths one may expect to discover SNe of both types and, among Type I's, approximately 1/3 to be of the anomalous subtype, just the same as it is found in spiral galaxies at present times. Therefore, at any z if dealing with SNe in spiral galaxies and at least for $z > 0.1$ in the case of the ellipticals one must make a *series* of accurate spectroscopic measurements to elucidate the nature of any given event. This, again, will limit the use of SNe as distance indicators to a redshift of $0.2 \div 0.3$, at most. In conclusion, supernovae seem to be suited to determine distances up to about 10^3 Mpc or $z \sim 0.1$ and, therefore, to determine the value of H_o accurately. However, in view of the practical limitations discussed above they can hardly be used for a direct determination of q_o.

REFERENCES

Arnett, W. D., 1982, *Ap. J.,* **253**, 785.

Barbon, R., Ciatti, F., Rosino, L., 1982, *Astr. Ap.,* **116**, 35.

Benvenuti, P., Sanz Fernandez de Cordoba, L., Wamsteker, W., Macchetto, F., Palumbo, G. C., Panagia, N., 1982, ESA SP-1046.

Benvenuti, P., Dopita, M., D'Odorico, S., 1984, private communication.

Bertola, F. 1964, *Ann. Astrophys.*, **27**, 319.

Bertola, F., Mammano, A., Perinotto, M., 1965, Contr. Asiago Obs. No. 174.

Bode, M. F., Evans, A. 1980, *Monthly Notices Roy. Astron. Soc.*, **193**, 21 p.

Canizares, C. R., Kriss, G. A., Feigelson, E. D., 1982, *Ap. J. (Letters),* **253**, L17.

Cavallo, G., Caraveo, P. A., Bignami, G. F., 1981, 17th International Cosmic Ray Conference, Paris, V. 9, XG 5.1-6, p. 80.

Chevalier, R. A., 1981, *Ap. J.,* **251**, 259.

Chevalier, R. A., 1984, *Ap. J. (Letters),* **285**, L63.

Clark, D. H., Stephenson, F. R., 1977, "Historical Supernovae," Pergamon Press (Oxford, England)

Dwek, E., A'Hearn, M. F., Becklin, E. E., Brown, R. H., Capps, R. W., Dinerstein, H. L., Gatley, I., Morrison, D., Telesco, C. M., Togunaka, A. T., Werner, M. W., Wynn-Williams, C. G., 1983, *Ap. J.,* **274**, 168.

Dwek, E., 1983, *Ap. J.,* **274**, 175.

Elias, J. H., Frogel, J. A. Hackwell, J. A., Persson, S. E. 1981, *Ap. J. (Letters),* **251**, L13.

Fransson, C., 1984, *Physica Scripta*, **T7**, 50.

Fransson, C., Benvenuti, P., Gordon, C., Hempe, K., Palumbo, G. G. C., Panagia, N., Reimers, D., Wamsteker, W., 1984, *Astr. Ap.*, **132**, 1.

Jenkins, E. B., Rodgers, A. W., Harding, P., Morton, D. C. and York, D. G., 1984, *Ap. J.*, **281**, 585.

Kirshner, R. P., Willner, S. P., Becklin, E. E., Neugebauer, G., Oke, J. B., 1973, *Ap. J. (Letters)*, **187**, L97.

Merrill, K. M. 1980, I.A.U. Circ. No. 3444.

Pacini, F., Salvati, M., 1981, *Ap. J. (Letters)*, **245**, L107.

Palumbo, G. G. C., Maccacaro, T., Panagia, N., Vettolani, G., Zamorani, G., 1981, *Ap. J.*, **247**, 484.

Panagia, N., 1980, Proc. 2nd European IUE Conference, ESA SP-157, p. XXVII–XXXI.

Panagia, N., 1982, Proc. 3rd European IUE Conference, ESA SP-176, p. 31.

Panagia, N., 1984a, *Physica Scripta*, **T7**, 15.

Panagia, N., 1984b, Proc. 4th European IUE Conference, ESA SP-218, p. 15-18.

Panagia, N., 1984c, in preparation.

Panagia, N., *et al.* 1980, *Monthly Notices Roy. Astron. Soc.*, **192**, 861.

Panagia, N. *et al.* 1984, in preparation.

Salvati, M., 1983, I.A.U. Symp. No. 101 "Supernova Remnants and their X-ray Emission," eds. J. Danziger and P. Gorenstein, D. Reidel (Dordrecht-Holland), p. 177–182.

Sandage, A. and Tammann, G. A. 1974, *Ap. J.*, **194**, 559.

Sandage, A., Tammann, G. A., 1975, *Ap. J.*, **196**, 313.

Shklovskii, I. S., 1981, *Sov. Astron. Lett.*, **7**, 263.

Sramek, R. A., Panagia, N., Weiler, K. W., 1984, *Ap. J. (Letters)*, **285**, L59.

Tammann, G. A., 1978, *Mem. S.A.It.*, **49**, 315.

Tammann, G. A., 1982, in "Supernovae: A Survey of Current Research," eds. M. J. Rees and R. J. Stoneham, D. Reidel, Dordrecht, Holland, p. 371–403.

Vaucouleurs, de, G., 1979, *Ap. J.*, **227**, 729.

Weiler, K. W., van der Hulst, J. M., Sramek, R. A., Panagia, N., 1981, *Ap. J. (Letters)*, **243**, L151.

Weiler, K. W., Sramek, R. A., van der Hulst, J. M., Panagia, N. 1982, in "Supernovae," eds. Rees, M. J. and Stoneham, R. J., D. Reidel (Dordrecht-Holland), p. 281.

Weiler, K. W., Sramek, R. A., van der Hulst, J. M., Panagia, N. 1983. I.A.U. Symp. No. 101. "Supernova Remnants and their X-ray Emission," eds. J. Danziger and P. Gorenstein, D. Reidel (Dordrecht-Holland), p. 171-176.

Weiler, K. W., Sramek, R. A., Panagia, N., van der Hulst, J. M., Salvati, M., 1984, in preparation.

SUPERNOVA OBSERVATIONS AT McDONALD OBSERVATORY

J. Craig Wheeler
University of Texas at Austin

Abstract

The programs to obtain high quality spectra and photometry of supernovae at McDonald Observatory are reviewed. Spectra of recent Type I supernovae in NGC 3227, NGC 3625, and NGC 4419 are compared with those of SN 1981b in NGC 4536 to quantitatively illustrate both the homogeneity of Type I spectra at similar epochs and the differences in detail which will serve as a probe of the physical processes in the explosions. Spectra of the recent supernova in NGC 0991 give for the first time quantitative confirmation of a spectrally homogeneous, but distinct subclass of Type I supernovae which appear to be less luminous and to have lower excitation at maximum light than classical Type I supernovae.

1. Introduction

The use of supernovae to measure the distance scale has been discussed for some time. Rapid advances are being made which will put this technique on an important new quantitative basis. There are three crucial components to this effort. The first, and most important, is a collection of high quality data. Another component is the evolutionary and hydrodynamical calculations which give physical credence to any result. The last component is the most exciting currently because its great potential is on the verge of being realized. This is the calculation of realistic model atmospheres which will provide the vital link between the dynamical models and the observations.

This paper discusses one of the components of this triad, the cooperative program at McDonald Observatory to obtain first rate spectra and photometry. In a companion paper (see Wheeler and Sutherland in this volume) we discuss some of the physical modeling being done to understand the origin of Type I supernovae. For developments in calculating model atmospheres for supernovae, see the papers by Branch, Harkness, and Hempe in this volume, and Hershkowitz, Linder, and Wagoner (1985).

A major aspect of the supernova program at McDonald observatory is the acquisition of spectra by interested observers using telescope time originally scheduled for other purposes. I serve as coordinator and cheerleader for this effort (but have yet to set foot on the mountain during a supernova run because of my well established reputation as a bad weather jinx). In the next section a summary of recent observations is given and some recent Type I spectra are presented, along with brand new data establishing a well defined class of peculiar Type I events. We also have begun a program of deep CCD photometry to follow light curves to very faint magnitudes. This program is discussed in the last section.

2. Spectra

During the furor over SN 1981b in NGC 4536, we attempted to reconfigure the telescope from Coude to Cassegrain in mid-run to accommodate the supernova. This proved to be too disruptive on all concerned (administrative observer, real observers, and mountain staff). We have established a pattern now in which IAU telegrams are monitored and the observer with the first available scheduled Cassegrain time volunteers or is beseeched to devote some of the previously scheduled time to obtaining spectra of the supernovae. This procedure has worked moderately well, as shown in Table 1.

There is no guarantee that the data will be acquired immediately after the announcement, and sometimes the wavelength coverage is not as complete as would be ideal, but luck sometimes finds a dedicated observer at the telescope when discovery is announced, and excellent spectra are obtained often enough to be very useful.

There are always a few supernovae going off in any season, although one can not know when or where in advance. To take advantage of this statistical likelihood and to follow up supernovae after discovery it is possible to schedule telescope time in advance. We have used this technique to a limited extent. Martin Gaskell used such dedicated time to get a good late-time (150 day) spectrum of SN 1983n in M 83. As we learn more about supernovae, and begin to identify specific phases which are crucial to the physical interpretation, we may have to return to the more onerous procedure of reconfiguring telescopes to catch brief, important phases.

Table 1 contains a number of points of interest. The single spectrum obtained by Ed Barker of SN 1983g in NGC 4753 near maximum light is of importance because the host galaxy is one of the I0 (or Irr II) type which, like ellipticals, seem to produce only Type I supernovae. Oemler and Tinsley (1979) argued that I0 galaxies produce Type I supernovae at a particularly rapid clip for their mass and luminosity and that they show signs of recent star formation. They concluded that Type I supernovae come from relative young massive stars, not old stars as suggested by their occurrence in ellipticals. We will examine this spectrum closely to see if it contains any systematic differences from other classical Type I spectra that could be related to the morphology of the host galaxy.

The event discovered on March 29, 1984 in NGC 3169 was a Type II with a

TABLE 1
Recent Supernova Spectra Obtained at McDonald Observatory

GALAXY SN-TYPE	DISCOVERY DATE	OBSERVER	DATE	EPOCH
NGC4753 83g I	4/4/83	Barker	4/12/83	+4
NGC4051 83i ?	5/11/83	Wills	5/17/83	~max
NGC5236(M83) 83n I(p)	7/14/83	Gaskell	2/28/84	+150
		Wills		
NGC3227 83v I	11/4/83	Gaskell	11/14,16/83	+10
		Wills	12/31/83	+60
		Wills/Netzer	2/2/84	+90
		Levreault/Garnett	2/4,6,8/84	+95
		Gaskell	2/28/84	+115
NGC3625 83w I	12/6/83	Brown	1/6/84	+30
			2/2,3/84	+55
		Levreault/Garnett	2/4,5/84	+60
		Shafter	3/30/84	+120
			4/1/84	
NGC4419 I	1/4/84	Brown	1/29/84	+25
			2/1/84	
		Wills/Netzer	2/2,3/84	+30
		Levreault/Garnett	2/6,9/84	+35
		Brown	3/24/84	+80
		Shafter	3/31/84	+85
MCG 9-19-19 featureless	2/1/84	Shafter	4/1/84	+60
NGC3169 II + spike	3/29/84	Gaskell	3/30,31/84	~max
			4/2/84	
		Wills	5/26,27/84	+30
NGC0991 I (p)	8/28/84	Levreault	8/30/84	~max
		Barker	9/21/84	+20
		Garnett/Dinerstein	10/19/84	+50
IC 121 I	8/29/84	Barker	9/22/84	+20

surprising new twist. In this case Martin Gaskell was on the 107 inch telescope when the discovery was announced and he got a spectrum which showed an unprecedented narrow emission spike on top of the usual broad H_a feature (IAU Circular 3936). He was able to resolve the narrow component to show that it had a FWHM of 430 km/s, thus proving that it was associated with the supernova, not a contaminating HII region. Pre-discovery plates show that the object brightened a year or two before the explosion (Gaskell and Keel 1985).

We have gotten fairly good coverage of three recent bright Type I events (~ 12 - 14^m at discovery) in NGC 3227, NGC 3625, and NGC 4419. Figures 1 to 3 show representative spectra of these events taken by two of our graduate students, Russell Levreault and Don Garnett. They are compared to spectra obtained at roughly the same epoch for SN 1981b. The fact that Type I spectra are very similar at the same epoch is frequently cited as the prime evidence that the underlying phenomena are very homogeneous. In the past, however, these claims were frequently based on comparison of spectra from modern electronic spectrographs with old photographic spectra, or with spectra presented in a journal on a different scale. With these data, all obtained with the same excellent equipment, we can plot the spectra on the same scale and manipulate them numerically to look for detailed quantitative similarities and differences. We can then attempt to reproduce the spectra with model atmospheres, to understand the physical differences, if any, between different Type I events.

Figure 1 gives the spectrum of the NGC 4419 event about 30 days after maximum light in the upper spectrum compared with the spectrum of SN 1981b in NGC 4536 at 29 days past maximum (assuming maximum light to be March 7, 1981). Note the overall similarity of the two spectra, both in amplitude and location of features. The prominent minimum at 6100 A is attributed to Si II. The absorption at 3800 A is thought to be due to Ca H and K, and that at 8200 A to the Ca infrared triplet. The features between 4000 and 6000 A are presumably blended Fe lines (Branch et al 1983).

A special effort was made to go as blue as possible with this spectrum of the 4419 event to check the reality of the purported Co II feature at 3200 A seen at maximum light in SN 1981b (Branch et al 1983). This feature is crucial to the interpretation of the deflagration models of Type I supernovae, because the stirring of heat from the inner, incinerated regions is necessary to obtain the partial burning of carbon and oxygen to produce their burning products, Si, S, Ca, identified in the spectra (Branch et al 1982, 1983). The stirring of heat also must mix the composition. In particular, according to detailed nucleosynthesis models (Nomoto, Thielemann, and Yokoi 1984; Woosley, Axelrod, and Weaver 1984; see also the papers by Branch and Harkness in this volume), there should be an appreciable amount of Co produced by the radioactive decay of Ni mixed in with the Ca.

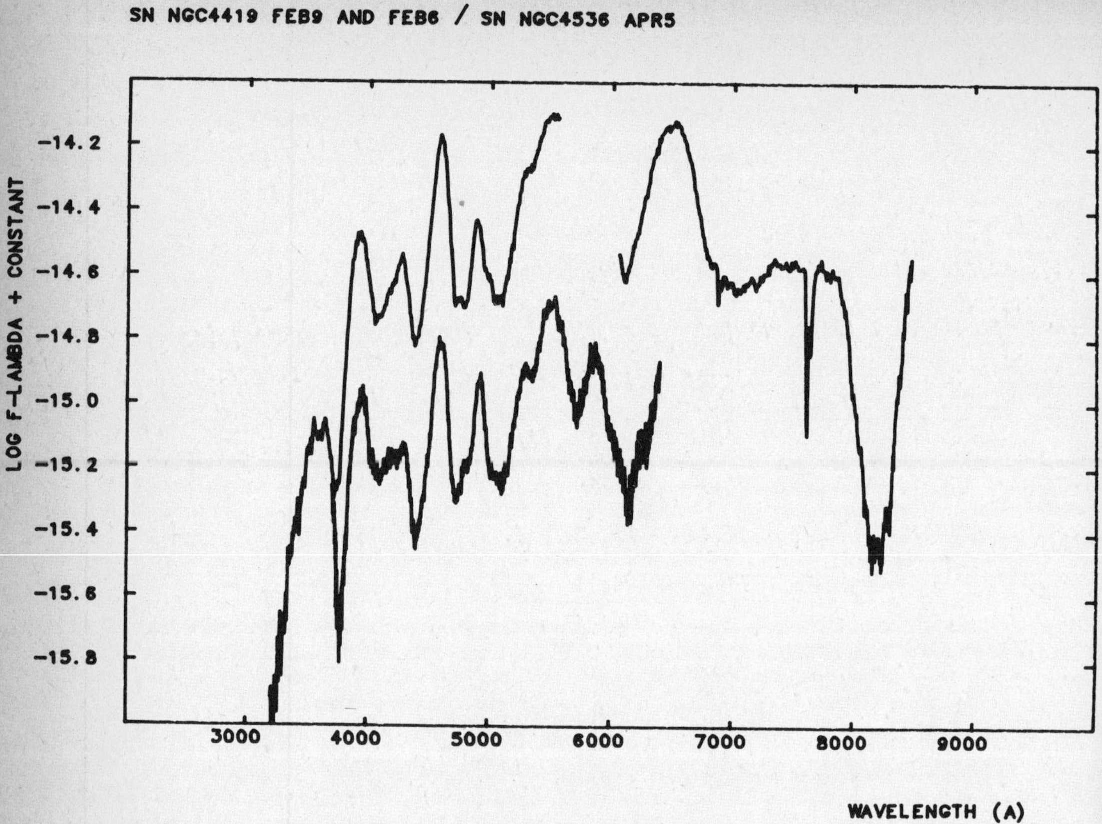

Figure 1. - The spectra of the supernova in NGC 4419 obtained by Russell Levreault and Don Garnett on February 6 (blue) and February 9 (red) 1984 approximately 30 days after maximum light (upper curves) are compared to the spectrum of SN 1981b 29 days after maximum light (lower curve, Branch et al 1983). Because the spectra were not necessarily taken under photometric conditions an arbitrary relative vertical shift has been applied to align the two spectra of the NGC 4419 event. The spectrum of the event in NGC 4419 has been corrected for a galactic radial velocity of -229 km/s, and the spectrum of SN 1981b for a radial velocity of 1809 km/s in accord with the Revised Shapley Ames Catalog.

Unfortunately, useful data is very difficult to obtain at the shortest wavelengths because of the atmospheric cutoff and technicalities associated with the reduction of the data. In the case of the 4419 event, appreciable signal was obtained down to 3000 A. Unfortunately, the last calibration point used to construct the spline fit to the comparison star was at 3200 A, so that the reduced data shortward of that is not trustworthy. This means the reliable data cuts off at 3200 A, right at the minimum of the suspect Co feature, so very little can be said about it. A close inspection of the extrapolation of the spline may enable the extraction of more information. The same problem of spline extrapolation may exist in the data of SN 1981b, so the status of the 3200 A absorption feature must be re-examined carefully there as well. The current status is that there is no evidence that the feature does not exist, a conclusion which would be an extreme blow to the deflagration model, but proof that it does exist is tantalizingly just beyond reach. Confirmation of this feature should be straightforward with the space telescope.

Figure 2 gives the spectrum of the Type I event in NGC 3625 approximately 60 days after maximum light in the upper spectrum and that of SN 1981b at 58 days (blue) and 64 days (red) after maximum light in the lower spectrum. Again, there are many startling quantitative similarities in the spectra, but some quantitative deviations in terms of the amplitude of various details. The most noticeable difference in the two spectra is the lack of the strong absorption at 6800 A in the 3625 event. This feature first appears in SN 1981b at about 24 days after maximum (Branch et al 1983) and persists through the end of the first season at 116 days (see the lower spectrum in Figure 3). It was the single feature which was not identified by Branch et al (1983) as being due to either Fe or Na in the later spectra. This feature does not seem to be present with significant strength in the second season observations of SN1981b (Branch 1984). The narrow features around 6200 A in the spectra of the 3625 event are of dubious reality since they fall right at the join between the February 4 and 5 spectra. Nevertheless, there may be some differences from SN 1981b in this region.

Figure 3 gives the spectrum of the Type I event in NGC 3227 obtained by Russell Levreault and Don Garnett about 90 days after maximum light compared with the spectrum of SN1981b 116 days after maximum light. Once again there are many similarities, but the absorption at 6800 A, while present, is not as strong as in the spectrum of SN 1981b.

The striking new piece of data which I have to present in this paper is the spectrum shown in Figure 4 obtained by Russell Levreault on August 28 and reduced only two days before the meeting. The spectrum is that of the supernova discovered in NGC 0991 by Evans (IAU Circular 3979) who estimates by visual

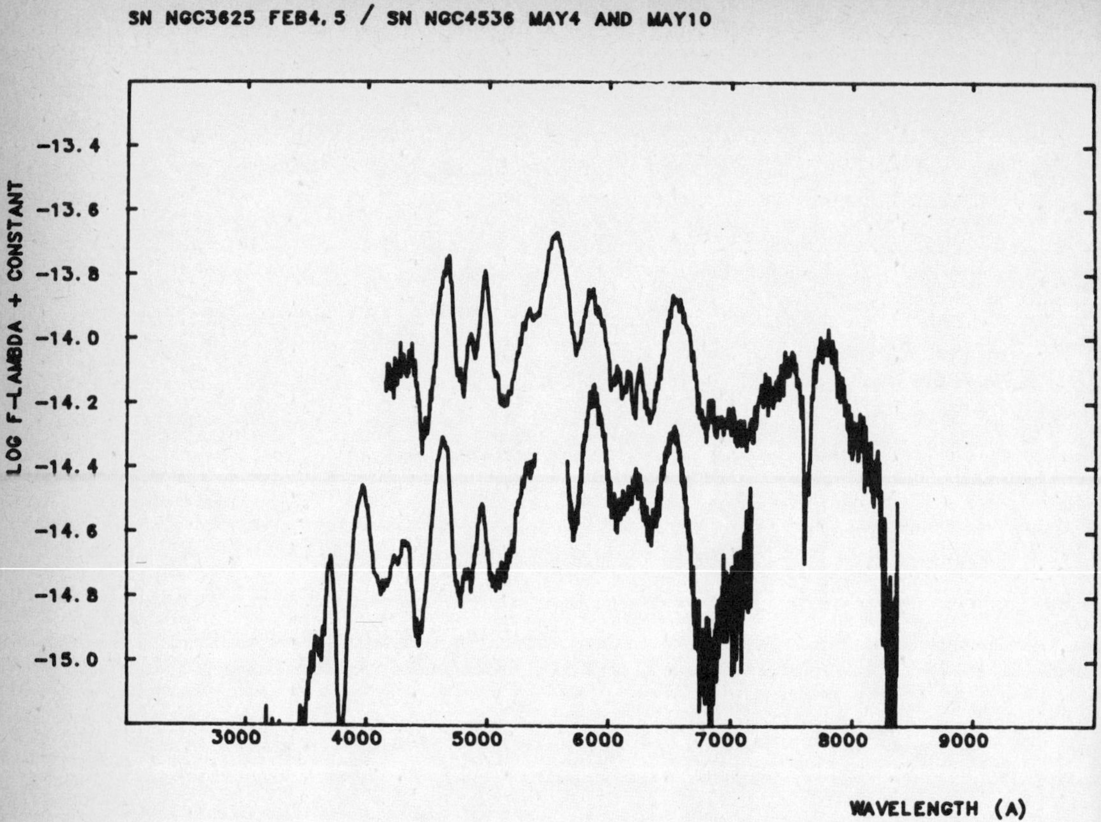

Figure 2. - Spectra of the Type I supernova in NGC 3625 obtained by Russell Levreault and Don Garnett on February 4 (blue) and February 5 (red) 1984 approximately 60 days after maximum light (upper curve) are compared to spectra of SN 1981b (lower curves) 58 days (blue) and 64 days (red) after maximum light (Branch et al 1983). The components of each of these pair of spectra are aligned with an arbitrary vertical shift (see caption of Figure 1 and text). The spectrum of SN 1981b has been corrected for a galactic radial velocity of 1809 km/s in accord with the Revised Shapley Ames Catalog, but the spectrum of the event in NGC 3625 is uncorrected in the absence of an available redshift.

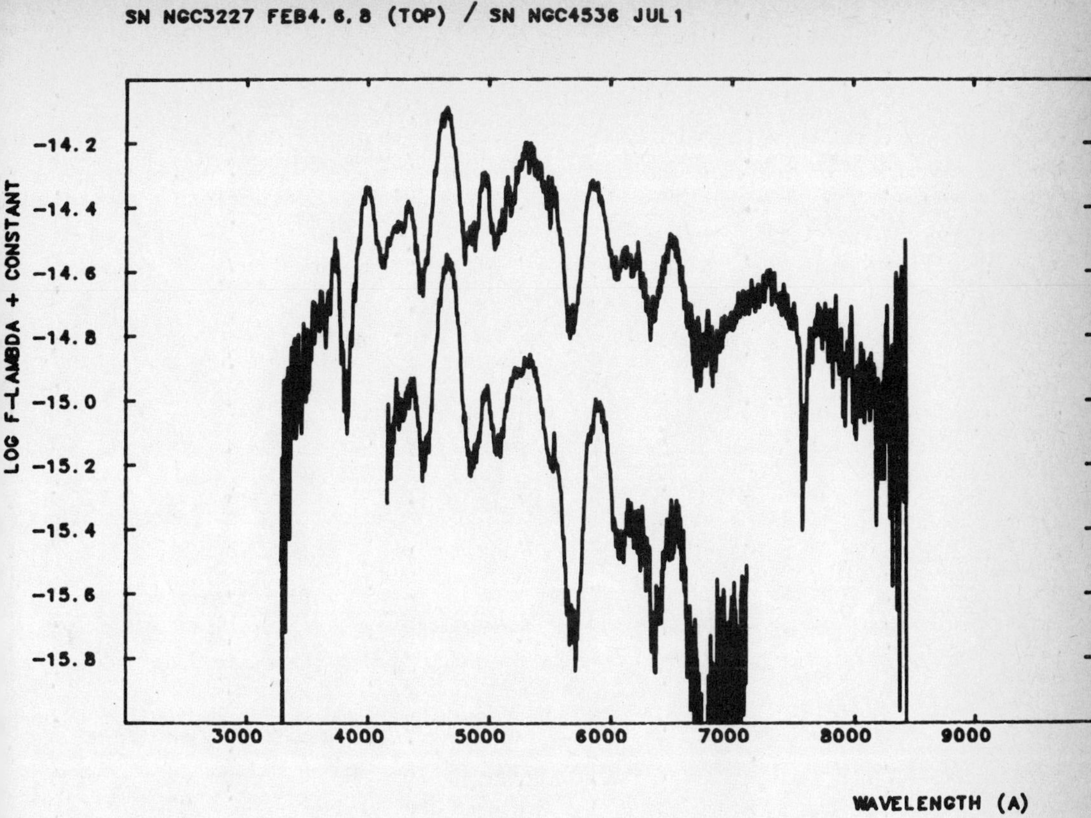

Figure 3. - Spectra of the Type I supernova in NGC 3227 obtained by Russell Levreault and Don Garnett on February 4 (blue) February 6 (yellow) and February 8 (red) 1984 approximately 90 days after maximum light (upper curve) are compared to a spectrum of SN 1981b (lower curve) 116 days after maximum light (Branch et al (1983), the spectrum of SN1981b 95 days after maximum light was not readily available). The components of the spectra of the 3227 event are aligned with an arbitrary vertical shift (see caption of Figure 1). The spectrum of the event in NGC 3227 has been corrected for a galactic radial velocity of 1198 km/s, and the spectrum of SN1981b for a radial velocity of 1809 km/s in accord with the Revised Shapley Ames Catalog.

observations that the supernova was near maximum light at $m_V \sim 13.2$ at the time the spectrum was obtained (IAU Circular 3989). Also shown for comparison are the spectra of SN 1981b at 58 days and 64 days past maximum reproduced from Figure 2. There is a difference in continuum slope between the two spectra, but striking wavelength coincidences between the two spectra for virtually all the features shortward of 6000 Å. The main difference in the spectra is the "doublet" at 6300 and 6500 Å in the 0991 event. There is a doublet of sorts in the SN 1981b spectra, but the minima are at 6100 and 6400 Å, distinctly different wavelengths, and in fact displaced so that the peaks in one spectrum are juxtaposed on the minima of the other. This feature and the fact that the maximum light spectrum otherwise resembles a post-maximum spectrum of SN 1981b thus represent a major departure from the spectra of classical Type I supernovae.

Given allowances for plotting on different scales, however, the spectrum of the 0991 event in Figure 4 is virtually identical to a spectrum of the supernova 1983n obtained near maximum light by Alan Uomoto of the University of Michigan (private communication, see also Richtler and Sadler 1983). Except for a relatively minor difference in continuum slope (the 0991 event is a little flatter), there is a one-to-one identification of the features throughout the spectrum, including the odd "doublet" at 6300 and 6500 Å. We find the evidence for the similarity between the two events compelling.

In classical Type I supernovae the principle identifying feature in the spectrum is the Si II λ 6355 feature which has a blue shifted absorption feature at about 6100 Å, and a broad emission peak centered at about 6400 Å. The odd "doublet" that characterizes SN 1983n and the 0991 event appears just where this Si emission should be. The Si absorption feature is conspicuous by its absence. A number of Type I supernova have been discovered in the past in which the Si feature appeared to be absent: SN 1962l in NGC 1073 (Bertola 1964), SN 1964l in NGC 3968 (Bertola, Mammano, and Perinotto 1965), and perhaps SN 1975b in an anonymous galaxy in the Perseus cluster (Kirshner, Arp, and Dunlap 1976). These spectra were photographic or variously contaminated, so a precise comparison was difficult. The striking quantitative similarity between the spectra of SN 1983n and the 0991 event, however, seem to establish that these events, and presumably the ones that preceded them, are members of a distinct and precisely defined subset of peculiar Type I supernovae. The spectral peculiarities of the 0991 event are probably responsible for its identification in various IAU circulars as a Type I at 9 to 17 and/or 20 to 60 days past maximum (Circular 3981), or a Type II about a month after maximum (Circular 3983).

SN 1983n was special for a number of reasons. As shown in Figure 4 the maximum light spectrum has strong similarities to Type I events after maximum light. For this reason Uomoto coined the phrase "born old" to describe SN 1983n. Since the

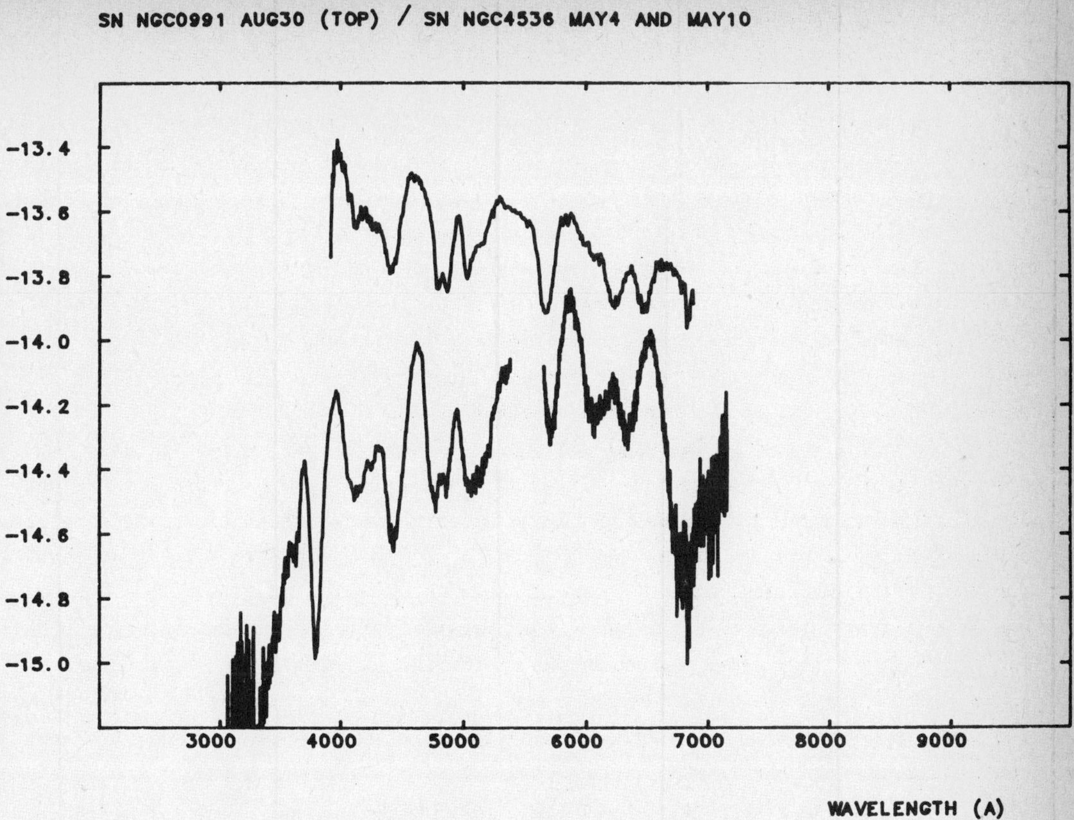

Figure 4. - A spectrum of the supernova in NGC 0991 obtained near maximum light by Russell Levreault on August 28, 1984, is compared to the spectra of SN1981b 58 (blue) and 64 (red) days after maximum light (Branch et al 1983). The spectrum of the 0991 event is corrected for a galactic radial velocity of 1530 km/s, and the spectrum of SN 1981b for a galactic velocity of 1809 km/s in accord with the Revised Shapley Ames Catalog. Although the spectra are very similar in the blue, the absence of the Si II absorption feature at 6100 A and the presence of the unidentified "doublet" at 6300 and 6500 A, and the close resemblance in just these features to the maximum light spectrum of SN 1983n establish that the 0991 event is a peculiar Type I supernova, and that this peculiar type represents a quantitatively well-defined subclass of Type I supernovae.

spectra of SN 1981b were qualitatively similar from 17 to 116 days after maximum light (Branch et al 1983), it would be too simplistic to identify the 0991 event with a particular phase of a classical Type I. To determine precisely how "old" the 0991 event is, i.e. to establish the composition, temperature, density, and velocity of the ejecta can only be done with a detailed comparison to synthetic spectra or supernova atmospheres.

The similarity of the spectrum of the 0991 event and the post-maximum spectra of SN 1981b in the blue and the reproduction of that portion of the SN 1981b spectrum by blended Fe II lines (Branch et al 1983) suggests that the 0991 event also contains iron traveling at velocities of about 10,000 km/s at the photosphere. The key to the difference of the peculiar Type I events is the absence of the Si feature and the presence of the odd "doublet".

The current spectra of SN 1983n and the 0991 event do not extend far enough to the blue to determine whether the distinctive Ca H and K line absorption feature at 3800 Å which marks classical Type I events is present. The data on the earlier peculiar Type I events hints that the Ca H and K absorption may be present (Branch, private communication), and infrared data apparently show that the Ca infrared triplet is present in SN 1983n (Uomoto, private communication). The spectra of the 0991 event obtained by Ed Barker on September 21 and by Don Garnett and Harriet Dinerstein on October 20 seem to show the characteristic Ca H and K dip in the unreduced data. A circumstance in which Ca was present, but Si absent would be most peculiar from a nucleosynthesis point of view, although a severely (and artificially) truncated deflagration might give such a result. Alternatively, the absence of high excitation Si II may represent an excitation, not a composition effect. An especially important task will be to identify the "doublet" by means of supernova atmosphere calculations.

SN 1983n was also conspicuous for being relatively dim at maximum light. At maximum it was about m_V ~ 11.5. Distance estimates for M 83 range up to 8 Mpc (Sandage and Tammann 1974). The peak absolute magnitude is thus dimmer than -18.5^m, whereas a classical Type I on this distance scale would be about -20^m. SN 1983n is thus of order 1.5^m dimmer than classical Type I events. NGC 0991 has a distance modulus of 29.03 (de Vaucouleurs private communication), and the supernova reached a peak apparent visual magnitude of about 13.2^m. By this measure, the 0991 event would have a magnitude of about -16^m whereas a classical Type I on this distance scale would be about -18.5^m. Thus the 0991 event also seems to be relatively dim by about the same amount as SN 1983n. This emphasizes that such peculiar Type I events must be identified by their spectra and removed from any determination which attempts to use Type I supernovae as standard candle distance indicators.

If the presence of Fe and the absence of Si at maximum light in the peculiar

Type I supernovae were to be interpreted as an abundance effect (ignoring the evidence for Ca for the moment) one might conclude that total incineration has occurred, leaving no intermediate mass elements. Various models exist in the literature involving complete detonation of a white dwarf, but these produce a great deal of Ni, and should be, if anything, brighter than a classical Type I which retains intermediate mass elements at maximum light. Alternatively, the lower peak luminosity suggests that the effective temperature is less at maximum light, given that the radius is constrained to be of order 10^{15} cm by the mass of the ejecta and the characteristic velocities. The differences could then plausibly be related to excitation, rather than abundance.

The lower luminosity may have a major impact on the model if the velocities are confirmed to be comparable to those in classical Type I events. The similarity between the spectrum of the 0991 event and that of SN 1981b in the blue (Figure 4) argues that the velocities are comparable, probably in excess of 8000 km/s. The obervations of SN 1983n over a period of about ten days prior to maximum light by Richtler and Sadler (1983) show a velocity shift of about 4000 km/s. This also implies the absolute velocities are comparable to those of classical Type I events.

If the light curves of peculiar Type I supernovae are also due to ^{56}Ni decay, then a lower luminosity may imply a small amount of Ni ejected. This might imply in turn that a thermonuclear explosion that produced only such a small amount of Ni could not account for the observed velocities (Sutherland and Wheeler 1984 , and Wheeler and Sutherland in these proceedings). If this proves to be the case, one may be forced to invoke another source of kinetic energy, such as core collapse, to account for the peculiar Type I supernovae. A physical understanding of these events clearly awaits a quantitative analysis of the spectra.

SN 1983n was also distinguished by being the first "Type I" supernova to be detected in the radio (see papers by Sramek and Weiler in these proceedings). Chevalier (these proceedings) has discussed the radio emission in the context of a shock propagating into a circumstellar nebula. Whether this picture applies only to these peculiar events, or to all Type I events is not clear. Certainly the event in NGC 0991 should be closely monitored for radio emission.

3. Long Term Photometry

Although the spectra contain the greatest concentration of information on a supernova, photometric light curves are also important ingredients in the study of supernova. In addition to being a useful guide to the phase when spectra are being obtained and the detectability of a supernova at a given epoch, light curves can be an important physical diagnostic, as well.

The most carefully worked out theory of Type I light curves (Axelrod

1980) argues that at late times, of order two years, the ejecta should cool sufficiently that the emissivity should shift rapidly from optical to infrared emission from iron fine structure lines. The available evidence from late time light curves (Barbon, Capellarro, and Turatto 1984) is sparse, but contains no hint of an "infrared catastrophe" , nor even of a significant departure from an exponential decline prior to 600 days. At 600 days Type I supernova are about 10^m below maximum. The task of obtaining photometry at these phases is difficult, but not impossible. Barbon et al. have also compiled data that suggest that some Type II supernovae may also undergo a late phase of linear decline, which may also be related to radioactive decay (Weaver and Woosley 1980). Monitoring Type II light curves would also be useful to add to the statistical sample of "linear" and "plateau" type light curves. Type II supernovae are usually fainter than Type I events, but the task of adding useful data on late time light curves is not so stringent, because few have been monitored more than 6^m below maximum.

A collaborative program involving Wheeler, Ed Robinson, and Marion Frueh, the staff observer at McDonald Observatory, has been initiated to routinely follow bright supernovae, $m_{max} < 15$, to very faint magnitudes, ~ 22 - 24, using a CCD. This is a long term program and we have no interesting results to report yet but we have made photometric observations of the following supernovae: NGC 3044, NGC 3169, NGC 3227, NGC 3625, NGC 4051, NGC 4220, NGC 4246, NGC 4419, NGC 4699, NGC 5236 (M 83), NGC 6217, NGC 6907, MCG 6-36-55, NGC 7184, IC 121, and NGC 0991.

In addition to this specific program, other observers contribute on a regular basis to the acquisition of photometric data on supernovae. Harold Corwin and another of our graduate students, Rick Binzel, have been particularly active in this regard.

Aknowledgments

This observational program would not exist without the active and enthusiastic participation of all the observers mentioned by name herein, and others who, for one reason or another, have not made contributions recently. Special thanks go to Harlan Smith who has provided strong support for the program despite the occasional disruptions it entails. I am particularly grateful to David Branch whose insights into the physical nature of supernova spectra and general enthusiasm helped to catalyze our efforts. My thanks to Robert Harkness for assembling the data and plotting the figures presented here. This research is supported in part by NSF Grant 8201210.

References

Axelrod, T. S. 1980, Ph.D. thesis, University of California, Santa Cruz.

Barbon, R., Cappellaro, E., and Turatto, M. 1984, preprint.

Bertola, F. 1964, Annals d'Astrophysique, **27**, 319.

Bertola, F., Mammano, A., and Perinotto, M. 1965, Contributions of the Asiago Observatory, #174.

Branch, D. 1984, in Proceedings of the Eleventh Texas Symposium on Relativistic Astrophysics, eds. D.S. Evans (New York: N.Y. Acad. of Science) , 186.

Branch, D., Buta, R., Falk, S.W., McCall, M.L., Sutherland, P.G., Uomoto, A., Wheeler, J.C., and Wills, B.J. 1982, Ap. J. Letters, **252**, L61.

Branch, D., Lacy, C. H., McCall, M. L., Sutherland, P. G., Uomoto, A., Wheeler, J. C., Wills, B. J. 1983, Ap. J., **270**, 123.

Gaskell, C. M., and Keel, W. C. 1985, in preparation.

Hershkowitz, S., Linder, E., and Wagoner, R. V. 1985, Ap. J., in press.

Kirshner, R. P., Arp, H. C., and Dunlap, J. R. 1976, Ap. J., **207**, 44.

Nomoto, K., Thielemann, F.-K., and Yokoi, K. 1984, in preparation.

Oemler, A., and Tinsley, B. M. 1979, A. J., **84**, 985.

Richtler, T., and Sadler, E. M. 1983, Astronomy and Astrophysics, **128**, L3.

Sandage, A., and Tammann, G. A. 1974, Ap. J., **194**, 559.

Weaver, T. A., and Woosley, S. E. 1980, Ann. NY Acad. Sci., **336**, 335.

Woosley, S.E., Axelrod, T.S., and Weaver, T.A. 1984, in Proceedings of the Erice Workshop on Stellar Nucleosynthesis, ed. C. Chiosi and A. Renzini, in press.

SPECTROPOLARIMETRY OF SUPERNOVAE

Marshall L. McCall
David Dunlap Observatory and University of Toronto

Abstract

A new technique is introduced for testing the validity of the spherical symmetry approximation of the Baade method for determining distances to supernovae. Lines with P Cygni profiles produced by resonance scattering in an expanding atmosphere with non-circular isophotes should show a linear polarization in excess of that in the continuum. Thus, spectropolarimetry offers a means of measuring the roundness of a supernova envelope and assessing the reliability of Baade method distances. So far, data has been obtained for two Type I supernovae. The data are discussed and interpreted in terms of a simple two-component atmosphere model for the polarization in P Cygni scattering profiles.

I. Introduction

Baade method distances to supernovae are particularly valuable to the determination of the extragalactic distance scale, because they are completely independent of other calibrators. Recent studies of supernovae in Virgo cluster galaxies (Branch, et al. 1981, 1983) have yielded distances about a factor of two higher than values derived from the de Vaucouleurs hierarchy (de Vaucouleurs, et al. 1981; Buta and Turner 1983). As a result, questions have been raised about the validity of the assumptions required to implement the Baade method (de Vaucouleurs, et al. 1981; Wagoner 1981). One of these assumptions, that of spherical symmetry, is the subject of this paper.

First a brief review of the approximations which go into the Baade method is given. The consequences of a breakdown in spherical symmetry are examined in particular detail. It is shown that the polarization profiles of the resonance scattered P Cygni lines are potentially a valuable probe of envelope asymmetries. Finally, the first spectropolarimetry of supernovae is presented. Initial interpretations are discussed using a simple two-component atmosphere model.

II. The Baade Method

To determine the distance to a supernova, the angular diameter estimated from observations of the optical flux and the colour temperature of the optical continuum is compared with the physical diameter determined from the P Cygni expansion velocity and time since outburst. To explore the effects of the spherical symmetry assumption, it will be assumed that a supernova envelope presents elliptical isophotes with an axis ratio (minor relative to major) $\mu(t)$, the orientation of which is fixed. The distance is given by

$$d^2 = \frac{\pi \, S_\lambda(T(t))}{F_\lambda(T)} \, \mu(t) \left[\alpha_o + V_\alpha(t) \, (t-t_o) \right]^2 \qquad (1)$$

where F_λ is the apparent flux at wavelength λ, $S_\lambda(T)$ is the surface brightness at colour temperature T, V_α is the velocity of expansion of the semimajor axis α, α_o is the semimajor axis of the progenitor, and t_o is the time of outburst.

For a geometrical asymmetry to persist to times much beyond the time of outburst, the expansion of the supernova must be anisotropic. To link the velocity of expansion along the major axis to the observed velocity, the actual geometry of the envelope must be known. For the purposes of discussion here, it is presumed that the envelope takes on the shape of an oblate spheroid. Then, the velocity field at any time must be given by

$$v = (r/a)\, V_{\alpha} \tag{2}$$

where v is the true velocity of expansion of matter at the photosphere at unprojected distance r from the centre of the supernova. If $\mu_\circ$ is the axis ratio as viewed by an observer in the equatorial plane (i.e. inclination $i = 90°$), the apparent velocity of expansion V_{app} (corresponding to the centres of the P Cygni absorption troughs), is given by

$$V_{app}(t) = g \frac{\mu_\circ}{\mu} V_{\alpha}(t) \tag{3}$$

where

$$\mu^2 = \cos^2 i + \mu_\circ^2 \sin^2 i \tag{4}$$

and g is a constant determined by the velocity field and the limb-darkening law. Substituting (3) in (1),

$$d^2 = \frac{\pi S_\lambda (T(t))}{F_\lambda (t)} \mu \left[\alpha_\circ + \frac{1}{g} \frac{\mu}{\mu_\circ} V_{app}(t)\,(t-t_\circ) \right]^2 \tag{5}$$

In past distance analyses, four assumptions have been made:

1. The size of the photosphere is much greater than the size of the progenitor star, i.e. $\alpha_\circ \ll \alpha$.

2. The density gradient in the envelope is sufficiently steep that the expansion velocity inferred from the P Cygni profiles is identical to the expansion velocity of material located at the photosphere,

and that $g \sim 1.0$.

3. The surface brightness of the optical continuum is that of a black body with a colour temperature equal to the colour temperature of the optical continuum, i.e. $S_\lambda(T) = B_\lambda(T)$ in the optical.

4. The outburst is spherically symmetric, i.e. $\mu = \mu_o = 1$.

The first assumption can be justified by comparing the envelope size with the estimated dimensions of supernova precursors (see Branch, et al. 1981), and the second assumption appears to be valid within the first 30 days after maximum light (Branch 1980; Branch, et al. 1981). The third assumption, which has generated the most controversy (de Vaucouleurs, et al. 1981; Wagoner 1981; Branch, et al. 1981), is discussed elsewhere in this volume.

The fourth assumption has been a subject of attention only recently (Shapiro and Sutherland 1982; McCall 1984; McCall, et al. 1984), primarily because observational evidence for asymmetries have been lacking. However, Tuohy, Clark, and Burton (1982) have found a young galactic supernova remnant which has a bright bar-like structure embedded in a faint extended halo. Also, two recent Type II supernovae (SN 1979c in M100 and SN 1980k in NGC 6946) appear to have been anomalously bright (Buta 1982). Theoretically, asymmetrical expansion could result if the progenitor were rapidly rotating or a member of a binary system. The former condition might apply to the progenitors of Type II supernovae, and the latter to the progenitors of supernovae of Type I. Asymmetries in nova shells have been attributed to interactions between the ejecta and the secondary during outburst (Hutchings, 1972). It seems reasonable to conclude that at least some supernova explosions may not be spherically symmetric. Therefore, the fourth approximation of the Baade method warrants serious investigation.

Adopting the first three asumptions, equation (5) becomes

$$d = \mu^{1/2} \left[\frac{\pi B_V (T(t))}{F_V (t)}\right]^{1/2} \left[\frac{\mu}{\mu_\circ} V_{app} (t)\right] (t-t_\circ) \qquad (6)$$

The factor $\mu^{1/2}$ arises from the distortion of the projected disc. The additional factor $\mu / \mu_\circ$ originates from the projection of an anisotropic velocity field (here that of an oblate spheroid). It can be seen that, regardless of the expansion law, a breakdown in the spherical symmetry approximation could lead to a distance too high by a factor $\mu^{-1/2}$. An error of a factor of two could arise if $\mu \leq 0.25$. However, it should be noted that the effect of the anisotropic expansion on distance determinations opposes the effect of the distortion. Therefore, it is possible also that the spherical symmetry approximation could lead to an underestimate of the distance. For the assumed velocity field, the probability of an overestimate ($71° < i < 109°$) is 32% for $\mu_\circ = 0.25$.

III. Polarization

1) Theory

In a supernova atmosphere, most lines appear to be produced by resonance scattering and the continuum by electron scattering above a photosphere emitting a thermal continuum (Branch 1980). Because scattering processes dominate the opacity, radiation emerging at an angle to the surface normal is linearly polarized. Therefore, if the atmosphere departs from spherical symmetry, a residual polarization may be observed (Harrington and Collins 1968; Shapiro and Sutherland 1982). Most research to date has concentrated on the continuum emerging from nonspherical atmospheres. Unfortunately, the continuum is not a useful probe of the symmetry of extragalactic supernova outbursts, because it is essentially impossible to determine whether any observed polarization is intrinsic to the supernova or a consequence of intervening dust.

The dust problem may be overcome by looking at the variation in polarization through the spectrum lines. For lines with P Cygni profiles, blueward absorption arises from the central regions of the disc as a result of scatterings out of the line of sight, and emission near the rest wavelength is produced by scatterings from the limb into the line of sight. If there is an asymmetry, the resulting linear polarization is expected to be higher in the lines than in the continuum. For the emission components, the increase comes from the addition of highly polarized radiation from the limb. For the absorption components, the increase results from the removal of unpolarized radiation from the centre of the disc.

Here, a simplified approach to the line problem is presented in order to make possible an initial assessment of the validity of the spherical symmetry approximation of the Baade method from recently acquired spectropolarimetric observations. It is hoped that this rough analysis will stimulate rigorous calculations of the polarization profiles for resonance scattered lines produced in nonspherical expanding atmospheres.

For opacities varying as r^{-n}, Cassinelli and Hummer (1971) found that for an extended pure scattering spherical atmosphere, the linear polarization of the continuum reaches a plateau

$$P_{\circ} = (n + 1) / (n + 3) \qquad (7)$$

at large radii where the tangential optical depth is much less than unity. The phase matrix for resonance scattering can be modelled as the sum of Rayleigh and isotropic scattering phase matrices (Chandrasekhar, 1960). The degree of depolarization caused by the isotropic scattering component depends upon the particular transition, but in general should lead to a plateau polarization smaller than that for the continuum.

Stimulated by the results of Cassinelli and Hummer, the disc of a supernova is approximated by a two-component atmosphere with elliptical isophotes having fixed apparent axis ratio μ. The disc is divided into a limb zone (designated by L) with constant surface brightness $I_{\lambda,L}$ at wavelength λ, from which radiation with polarization $P_\circ$ (either for a line or the continuum) emerges, and a central zone (designated by C) with constant surface brightness $I_{\lambda,C}$ from which emergent radiation is assumed to be unpolarized. The ratio of the apparent major axis of the limb zone to the apparent major axis of the central zone, designated by k, is set by the net polarization of the continuum radiation from the whole disc. This model will be used to estimate the net polarization of the total radiation in the emission and absorption components of the P Cygni profiles in terms of the polarization in the continuum.

After integrating the Stokes parameters for the limb zone, the net linear polarization of the radiation from the limb is calculated to be

$$P_L = P_\circ (1 - \mu) / (1 + \mu) \tag{8}$$

The factor k has cancelled out, because the shapes of zones C and L were assumed to be identical. The plane of polarization is aligned with the minor axis of the disc.

At an arbitrary wavelength, radiation may be contributed by both continuum and lines. However, in the two-component model, polarized radiation comes from the limb only. The total intensity of polarized radiation coming from the whole disc is

$$\ell_{\lambda,pol} = \pi\alpha^2 \mu (k^2-1) \left[P_L(\text{line})\, I_{\lambda,L}(\text{line}) + P_L(\text{cont})\, I_{\lambda,L}(\text{cont}) \right] \tag{9}$$

where P_L(line) and P_L(cont) designate the net polarizations of line and continuum radiation, respectively, coming from the limb zone. The total intensity of all radiation is

$$\ell_\lambda = \pi\alpha^2\mu(k^2-1) \left[I_{\lambda,L}(\text{line}) + I_{\lambda,L}(\text{cont}) \right] + \pi\alpha^2\mu \left[I_{\lambda,c}(\text{line}) + I_{\lambda,c}(\text{cont}) \right] \tag{10}$$

The polarization at wavelength λ is given by $\ell_{\lambda,pol}/\ell_\lambda$.

In the continuum, $I_{\lambda,L}(\text{line}) \sim 0$, and $I_{\lambda,c}(\text{line}) \sim 0$. The resulting expression for P(cont) contains the atmospheric extension factor k and the limb darkening law $I_{\lambda,c}(\text{cont})/I_{\lambda,L}(\text{cont})$, neither of which is known. While the model cannot predict P(cont), it can be used to estimate the polarization in the lines in terms of P(cont).

First consider a P Cygni absorption trough. In this case, $I_{\lambda,L}(\text{line}) \sim 0$, and $I_{\lambda,c}(\text{line})$ is modelled by a negative number. The ratio of the depth of the absorption to the level of the continuum is given roughly by

$$f(\text{abs}) = -I_{\lambda,c}(\text{line}) / \left[(k^2-1)\, I_{\lambda,L}(\text{cont}) + I_{\lambda,c}(\text{cont})\right] \quad (11)$$

Thus, the polarization in the absorption is

$$P\,(\text{abs}) = P\,(\text{cont}) \,/ \left[1 - f\,(\text{abs})\right] \quad (12)$$

At a wavelength located in a P Cygni emission peak, $I_{\lambda,c}(\text{line}) \sim 0$. The ratio of the height of the emission (above the continuum) to the level of the continuum is given roughly by

$$f(\text{emis}) = (k^2-1)\, I_{\lambda,L}(\text{line}) / \left[(k^2-1)\, I_{\lambda,L}(\text{cont}) + I_{\lambda,c}(\text{cont})\right] \quad (13)$$

so that the polarization in the emission is

$$P(\text{emis}) = \left[P(\text{cont}) + f(\text{emis})\, P_L(\text{line})\right] / \left[1 + f(\text{emis})\right] \quad (14)$$

The value of $P_L(\text{line})$ can be computed from (8) provided that $P_\circ$ is reduced appropriately to account for the depolarization for the particular transition.

From equations (12) and (14), it can be seen that the polarization in both the emission and absorption components of P Cygni profiles should exceed that in the continuum if the apparent disc is noncircular. As an example, consider a transition $J_1 = 0$, $\Delta J = +1$. If $n \sim 7$ (Branch 1980), then $P_\circ(\text{cont}) \sim 0.80$, so that $P_\circ(\text{line}) \sim 0.34$. For $P(\text{cont}) \sim 0.01$, $f(\text{abs}) \sim 0.5$, and $f(\text{emis}) \sim 0.25$, an axis ratio $\mu = 0.25$ would lead to $P(\text{abs}) \sim 0.02$ and $P(\text{emis}) \sim 0.05$. Such a contrast between emission and continuum polarizations should be readily measurable.

Any polarization that intervening dust introduces should vary only slowly with wavelength. While the shape of P Cygni polarization profiles may be altered (even inverted, depending on the difference between the intrinsic position angle and that of the interstellar polarization), the "discontinuity" between continuum and line polarizations should still reach the observer. Therefore, a change in polarization at the position of a P Cygni profile could be unambiguously interpreted as arising from an asymmetry in the supernova envelope.

2) Observations

Currently, because of the rarity of spectropolarimeters, their confinement to large telescopes, and the transience of the supernova phenomenon, it is exceedingly difficult to obtain spectropolarimetry of supernovae. Despite the obstacles, the author has had the good fortune of being in the right place at the right time on two occasions, which led to the acquisition of the first such observations. The data could not have been obtained or reduced without the consent and help of a great many people. Their names are acknowledged at the end of this paper.

Both times, observations were made with the Anglo-Australian Observatory Pockels Cell Spectropolarimeter (McLean, et al. 1984) with the RGO spectrograph and IPCS detector on the 4 meter Anglo-Australian Telescope. This device is particularly well suited for spectropolarimetry, because flat fields are eliminated as a source of error, and because the switching rate of the Pockels cell is fast enough that the effects of temporal variations in atmospheric transparency or instrumental response cancel out. Details concerning observations and reductions are given by McCall, et al. (1984).

The first supernova to be observed was SN1983g in NGC 4753, which is an IO galaxy in the south-east corner of the Virgo X Cloud. The observation was made on April 7, 1983, only one night after the discovery by R. Evans (see IAU Circ. No. 3789). The supernova was subsequently identified to be of Type I, at a phase close to maximum light (see McCall, et al. 1984).

The results of the spectropolarimetry are displayed in Figure 1. Although the supernova light is linearly polarized, with the mean level being about 0.02, no significant change in the polarization parameters is evident at the positions of the P Cygni profiles. Still, the data can be used to place a lower limit on the apparent axis ratio of the disc.

It is possible to use the two-component model to express the axis ratio in terms of measurable parameters. Combining equations (8) and (14),

$$\mu = \frac{f(\mathrm{emis})\,[P_o(\mathrm{line})-P(\mathrm{emis})] - [P(\mathrm{emis})-P(\mathrm{cont})]}{f(\mathrm{emis})\,[P_o(\mathrm{line})+P(\mathrm{emis})] + [P(\mathrm{emis})-P(\mathrm{cont})]} \qquad (15)$$

The axis ratio is constrained most severely by the lines of [Ca II]λ3934 and Si IIλ5041 ($J_1 = 1/2$, $\Delta J = 0$), for which the limb polarization P_o is highest. After accounting for blended lines, $P_o(\mathrm{line}) \sim 0.3 P_o(\mathrm{cont}) \sim 0.25$, assuming $n = 7$ (Branch 1980). From the Si II$\lambda\lambda$6347,6371 blend (not displayed), which should have a profile similar to that of Ca II$\lambda\lambda$3934,3968, it is estimated that $f(\mathrm{emis}) \sim 0.25$.

If the polarization introduced by intervening dust greatly exceeds the intrinsic polarization at all wavelengths, then $P(\mathrm{cont}) \sim 0$, so that $P(\mathrm{emis}) < 0.015$ and $P(\mathrm{emis})-P(\mathrm{cont}) < 0.015$, implying $\mu > 0.54$. On the other hand, if the intrinsic polarization of the supernova dominates at all wavelengths over the polarization introduced by the dust, then $P(\mathrm{cont}) \sim 0.015$. Consideration of the rms noise leads to 2σ upper

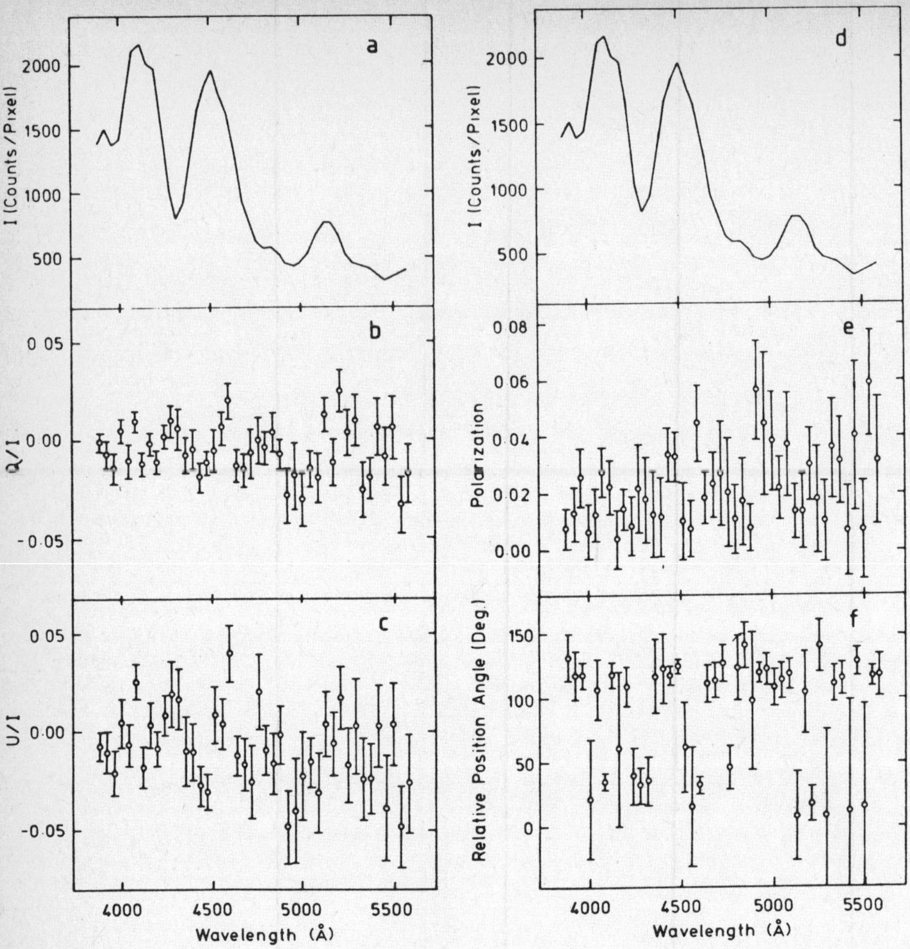

Figure 1. Spectropolarimetry of SN1983g.
(a),(d) The raw unfiltered spectrum. The ordinates are averages over 40 Å intervals. (b),(c) Q/I and U/I, averaged over 20 pixel (40 Å) intervals. (e) The linear polarization. (f) The position angle of the plane of polarization. The angles are relative, since the zero point was not calibrated. A comparison of panels (d) and (e) reveals no significant polarization excess in the P Cygni emission peaks.

limits of 0.03 for P(emis) and 0.15 for P(emis)-P(cont), implying $\mu > 0.47$. It is concluded that $\mu > 0.5$, and that application of the Baade method under the assumption of spherical symmetry could not overestimate the distance by more than a factor of 1.4.

The second supernova observed was SN1983n, which was located in the Sc galaxy M83 (NGC 5236) in the Centaurus group. Spectropolarimetric data between 3700 and 5200 Å was acquired on July 18, 1983, fifteen nights after the discovery (again by R. Evans), but within one night of maximum light (Sramek, Panagia, and Weiler 1984). Although the analysis of the data is not yet complete, preliminary results are exciting.

Optical and ultraviolet observations indicated that the supernova was of Type I, but peculiar and underluminous (Sramek, Panagia, and Weiler 1984). The peculiarity was further manifested by the presence of strong radio emission (Sramek, Panagia, and Weiler 1984). Although the observation of July 18 was made close to maximum light, the spectrum most closely resembles that of typical Type I supernovae about 20 days past maximum (see Branch, et al. 1983). The most prominent feature is a blend of Fe II lines, whose emission peak is centred around 4600 A.

Preliminary reductions indicate that the polarization spectrum dips from about 0.014 to about 0.008 at the position of the Fe II peak. No significant change in position angle is observed. Both of these observations can be explained if the interstellar polarization vector and the intrinsic polarization vector are close to being orthogonal.

If the "discontinuity" survives further scrutiny, it will be the first direct evidence for the existence of asymmetric supernova explosions. It is tempting to speculate that the other peculiarities of SN1983n are a consequence of a high inclination.

IV. Summary

Recent observations of supernovae and remnants have stimulated questions about the validity of the spherical symmetry assumption of the Baade method for determining distances. If this approximation were incorrect, conventional Baade method distances could be in error. The most effective way of checking the spherical symmetry of supernovae is by mapping the polarization across a resonance scattered P Cygni line profile. Regardless of the instrumental polarization and the polarization introduced by intervening dust, a change in polarization at the position of a line could be unambiguously interpreted as arising from an asymmetry in the supernova envelope.

Acknowledgements

The spectropolarimetric observations discussed here could not have been acquired or reduced without the help of the following people:

J. Bailey	Anglo-Australian Observatory
M. Bessell	Mount Stromlo and Siding Spring Observatories
D. Morton	Anglo-Australian Observatory
N. Reid	Royal Greenwich Observatory
D. Wickramasinghe	Australian National University
B. Vaile	University of New South Wales

The author is very grateful to all. Thanks are also conveyed to the Natural Sciences and Engineering Research Council of Canada for its continuing support.

References

Branch, D. 1980, in **Supernova Spectra**, ed. R. E. Meyerott and G. H. Gillespie (New York: American Institute of Physics), p. 39.

Branch, D., Falk, S. W., McCall, M. L., Rybski, P., Uomoto, A. K., and Wills, B. J. 1981, Ap. J., 244, 780.

Branch, D., Lacy, C. H., McCall, M. L., Sutherland, P. G., Uomoto, A., Wheeler, J. C., and Wills, B. J. 1983, Ap. J., 270, 123.

Buta, R. J. 1982, Pub. A.S.P., 94, 578.

Buta, R. J., and Turner, A. 1983, Pub. A.S.P., 95, 72.

Cassinelli, J. P., and Hummer, D. G. 1971, M.N.R.A.S., 154, 9.

Chandrasekhar, S. 1960, Radiative Transfer (New York: Dover).

de Vaucouleurs, G., de Vaucouleurs, A., Buta, R., Ables, H. D., and Hewitt, A. V. 1981, Pub. A.S.P., 93, 36.

Harrington, J. P., and Collins, G. W. II 1968, Ap. J., 151, 1051.

Hutchings, J. B. 1972, M.N.R.A.S., 158, 177.

McCall, M. L. 1984, M.N.R.A.S., in press.

McCall, M. L., Reid, N., Bessell, M. S., and Wickramasinghe, D. 1984, M.N.R.A.S., in press.

McLean, I. S., Heathcote, S. R., Paterson, M. J., Fordham, J., and Shortridge, K. 1984, M.N.R.A.S., 209, 655.

Shapiro, P. R., and Sutherland, P. G. 1982, Ap. J., 263, 902.

Sramek, R. A., Panagia, N., and Weiler, K. W. 1984, preprint.

Tuohy, I. R., Clark, D. H., and Burton, W. M. 1982, Ap. J., 260, L65.

Wagoner, R. V. 1981, Ap. J., 250, L65.

RADIO EMISSION FROM A TYPE I SUPERNOVA

R. A. SRAMEK
National Radio Astronomy Observatory*
Socorro, New Mexico

Over the last several years there have been several detections of radio emission from optically discovered supernovae of Type II (SNII). The luminosity at 6cm wavelength of these radio supernovae (RSN) can be as high as 300 times that of the powerful galactic supernova remnant Cas A. The properties of the several Type II RSN are reviewed in a paper by K. Weiler in these proceedings and by Weiler et al. (1985). However, until recently, there had been no detections of radio emission from Type I supernovae, leading to the speculation that this might be a fundamental distinction between Types I and II. The picture changed dramatically in July 1983 when strong radio emission was detected from the Type I supernova that appeared in M83 (Sramek, Panagia, and Weiler, 1984).

The supernova SN1983n (called SN1983.51 in the date of discovery convention) was first detected optically on July 3, 1983. Radio observations made on July 6 at the VLA gave a 6cm flux density of 2.0 mJy at the position of the supernova where earlier maps by J.M. van der Hulst (private communication) showed only a three sigma upper limit of < 0.15 mJy. Subsequent monitoring at the VLA gave a detailed light curve at 6cm and more sparse data at 20cm wavelengths. The observed data are plotted in Figure 1.

The radio emission is clearly non-thermal with an average spectral index of -1.04 during the period when both 6cm and 20cm data were available. The linear decline after day 30 on the log-log plot in Figure 1 indicates that the radio flux density has a power law decay in time. However, the first two 6cm data points deviate significantly from the later power law decay suggesting early absorption.

Following the model for radio emission from supernovae developed by Chevalier (1982,1984) the radio emission arises in the interaction region between the rapidly expanding supernova gas and the circumstellar envelope of slowly moving gas that was created by the presupernova stellar wind. This model predicts a power law decay in time after a period when the non-thermal emission is free-free absorbed by the

* The National Radio Astronomy Observatory is operated by Associated Universities, Inc., under contract with the National Science Foundation.

material in the circumstellar shell. In this model, the exponent of the power law time decay, b, is given by b = $-(6-2\alpha-6m)/2$, where α is the observed radio spectral index and m is related to the density profile of the expanding supernova envelope; for $\alpha=-1.0$ and m=0.8 (Chevalier, 1984), the theory predicts b= -1.6 .

The best fit curve to the 6cm data is given by

$$S(6cm) = 4263\ (t-t0)^{b}\ \exp(-T)\ \mathrm{mJy},$$

with b = -1.595 and the optical depth $T = 484.6(t-t0)^{-2.4}$. Here (t-t0) is the number of days since the supernova explosion, with t0 = 29 June 1983. The agreement with the predicted value for b is extremely good, offering strong support for the Chevalier model. The peak radio emission of the best fit light curve occurs near day 16 with a peak flux of 27.4 mJy. This is two days before the optical maximum, which may pose a problem for detecting future SNI.

The power law decay of this SNI is much faster than that of typical SNII where b ranges between -0.6 and -0.9 and where (in agreement with the above model) the radio spectra are flatter with indices of about -0.6 . It is uncertain why SNI generate relativistic electrons with steeper electron indices than SNII.

The progenitor of SN1983.51 was probably not a single isolated star, but rather a member of a binary system of two quite massive stars (see Sramek,Panagia, and Weiler 1984 and references therein). The observed free-free optical depth and the energy requirements for the radio emission set constraints on the circumstellar envelope. In particular, a mass loss rate of about 2.7 x 10^{-6} M/yr is suggested, which is the rate for a 6.5 M red supergiant. This supergiant was probably the companion of the supernova progenitor and supplied a circumstellar envelope about both stars in the binary system. Since the supernova progenitor evolved faster and thus was the more massive of the two stars, it was presumably almost massive enough (about 8 M) to become a SNII. In this scenario, radio emission from a SNI requires an exceptionally massive progenitor binary star system and suggests that detectable SNI radio emission may be quite rare.

I wish to acknowledge my collaborators in the study of radio supernovae for there boundless enthusiasm, energy, and insight: Kurt Weiler, Thys van der Hulst, and Nino Panagia.

References:

Chevalier,R.A., 1982, Ap. J., 259, 302

Chevalier,R.A., 1984, Ap. J. Lett., in press

Sramek,R.A., Panagia,N., and Weiler,K.W., 1984, Ap. J. Lett., in press

Weiler,K.W., Sramek,R.A., Panagia,N., van der Hulst,J.M., 1985, in preparation

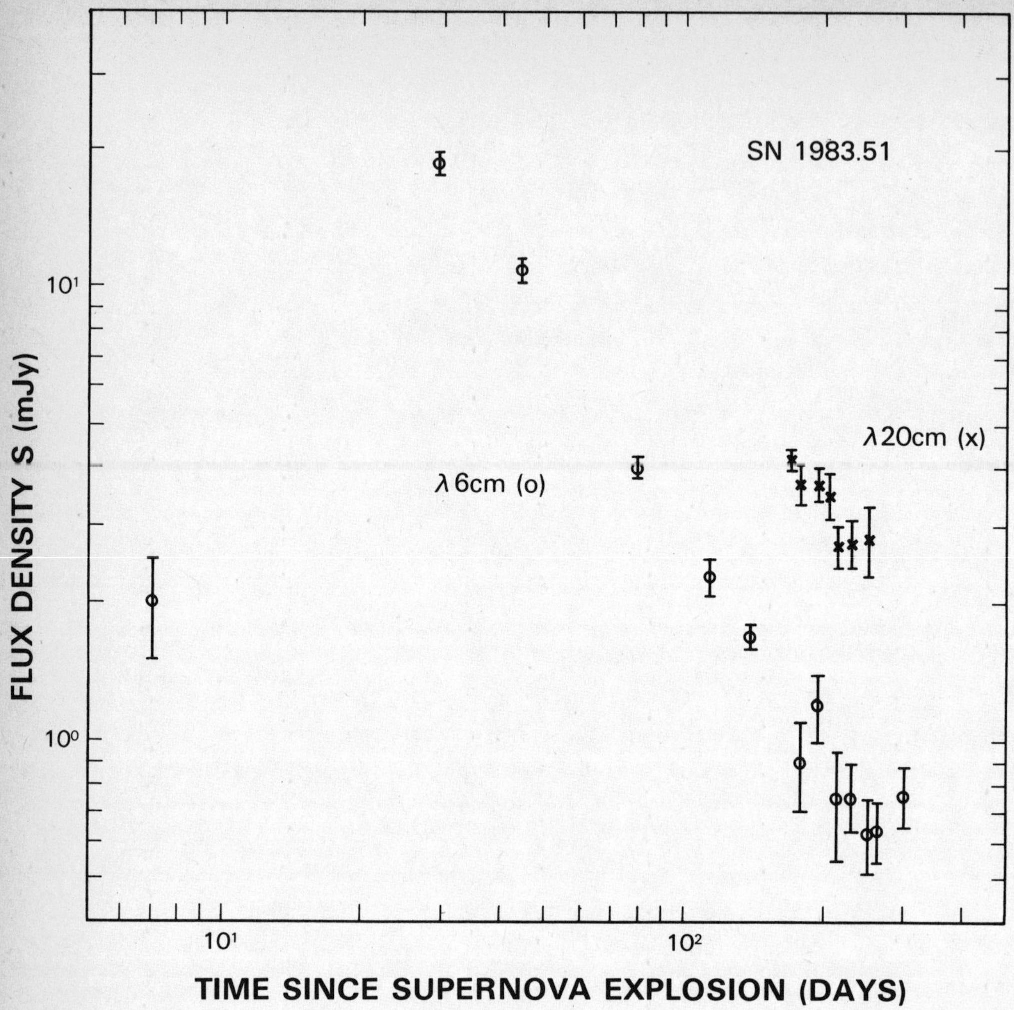

Figure 1. The radio light curve of the Type I supernova SN1983.51 (SN1983n). The epoch of the explosion is taken to be June 29, 1983. After day 30, the light curve is well fit by a power law with an index of -1.6.

RADIO EMISSION FROM TYPE II SUPERNOVAE

Kurt W. Weiler
National Science Foundation, Washington, D. C. 20550, U.S.A.

Over the past five years the study of radio emission from supernovae has developed from a few weak detections of SN1970g into a field with detailed, multifrequency radio "light curves" on several objects, with extensive spectral information, and with complex models for the radio supernova (RSN) phenomenon. This work is continuing and will provide within the next few years detailed studies of individual supernovae as well as statistics concerning the properties of the radio emission from different classes of supernovae.

I. INTRODUCTION

Numerous searches for radio emission from supernovae (de Bruyn, 1973; Brown and Marscher, 1978; Ulmer *et al.*, 1980) failed to detect any of several tens of objects which had been identified as optical supernovae during the last century. The only detections before 1980 were the results on 1970g (Gottesman *et al.*, 1972) which yielded in total four significant points at two wavelengths (Allen *et al.*, 1976).

II. RESULTS

When the bright Type II (subclass "L" or "linear") supernova SN1979c was discovered in M100 (NGC4321) in 1979 April, we turned the Very Large Array (VLA)[1] to trying to detect radio emission from the object. An initial measurement in 1979 April at 6 cm wavelength was unsuccessful but a year later in 1980 April, the VLA was again used at 6 cm to study the region and this time a source of 5 mJy, the brightest source in the radio map at that time, was found at the position of the optical supernova.

After this initial detection in 1980 April, we initiated monitoring at the two VLA wavelengths of 20 cm and 6 cm at a rate of approximately once per month with more infrequent observations at 2 cm. These observations have yielded the detailed radio "light curves" shown in Figure 1.

As can be seen from Figure 1, the radiation was initially optically thick at 20 cm. Later, the source also "turned-on" at 20 cm and has by now become optically thin at both 6 and 20 cm with a final spectral index of $\alpha \sim -0.6$. The change of the spectral index between 6 and 20 cm for SN1979c with time is shown in Figure 2.

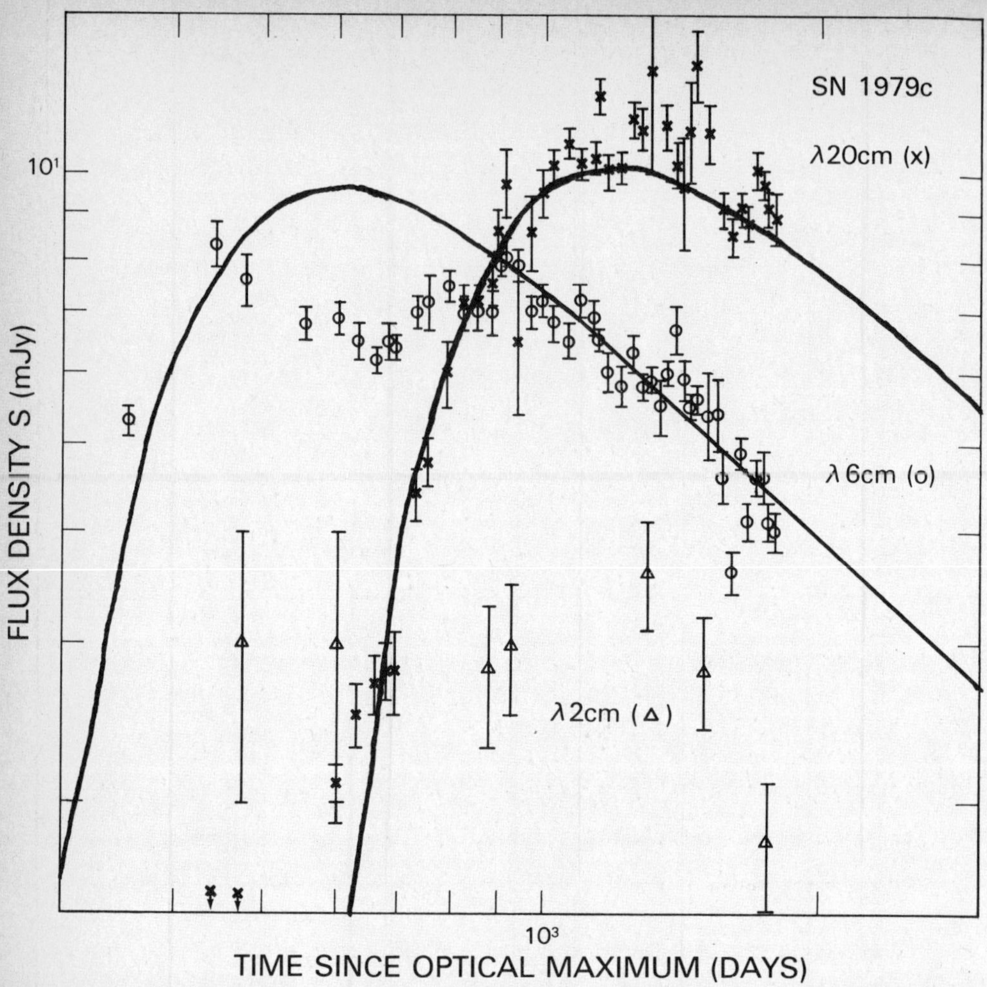

Figure 1: Radio light curves for the Type II supernova SN1979c in NGC4321 (M100). All three wavelengths measured with the VLA are shown together; 20 cm (x), 6 cm (o), and 2 cm (Δ). The "age" of the supernova is measured in days from the date of maximum optical light on 1979 April 19. The solid lines represent the "best" fit of curves of the form $S = K_1 \nu^{\alpha} t^{\beta} e^{-\tau}$ where $\tau = K_2 \nu^{-2} t^{\delta}$ with $K_1 = 1.07*10^4$, $\alpha = -0.56$, $\beta = -0.92$, $K_2 = 4.31*10^{10}$, and $\delta = -3.54$.

In addition to the detailed studies of SN1979c, we had the good fortune to have a second Type II supernova, also of the subclass "L," appear in the galaxy NGC6946 in 1980 November. Again our radio monitoring was started while the object was still optically bright and this time a year long wait for 6 cm radio emission was not necessary. At an age after maximum optical light of only 35 days, a radio source was detected at 6 cm wavelength at the position of the supernova.

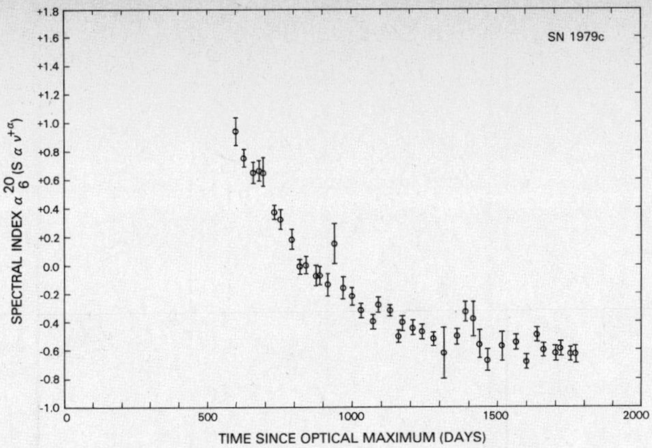

Figure 2. Spectral index (α) for SN1979c between 20 cm and 6 cm ($S \propto \nu^{+\alpha}$) plotted as a function of time in days since maximum optical light on 1979 April 19.

In much the same fashion as for SN1979c regular, approximately monthly monitoring of the supernova at both 6 and 20 cm, with less frequent observation at 2 cm, was initiated with the VLA. Detailed light curves have been established for SN1980k and these are shown in Figure 3.

The spectral index change with time for SN1980k is shown in Figure 4 in a manner similar to that for SN1979c in Figure 2.

In summary, we know that at least SN1979c has low linear polarization, being less than 1% polarized at 6 cm wavelength, that both supernovae have fast turn-on times at each wavelength followed by slow declines with time, that they are optically thin first at shorter wavelengths and later at longer wavelengths, and that they are nonthermal emitters having both high brightness temperatures and nonthermal spectral indices.

III. MODELS

In order to try to understand the extensive data sets available for these two Type II supernovae and to relate them to the question of this conference, supernovae as distance indicators, it is necessary to consider the models for their radio emission. Several have been proposed. The simplest and oldest is that of an expanding relativistic gas cloud created essentially instantaneously by an explosion. This balloon of relativistic particles and magnetic fields then expands freely and adiabatically. This model was not developed for supernovae, but for

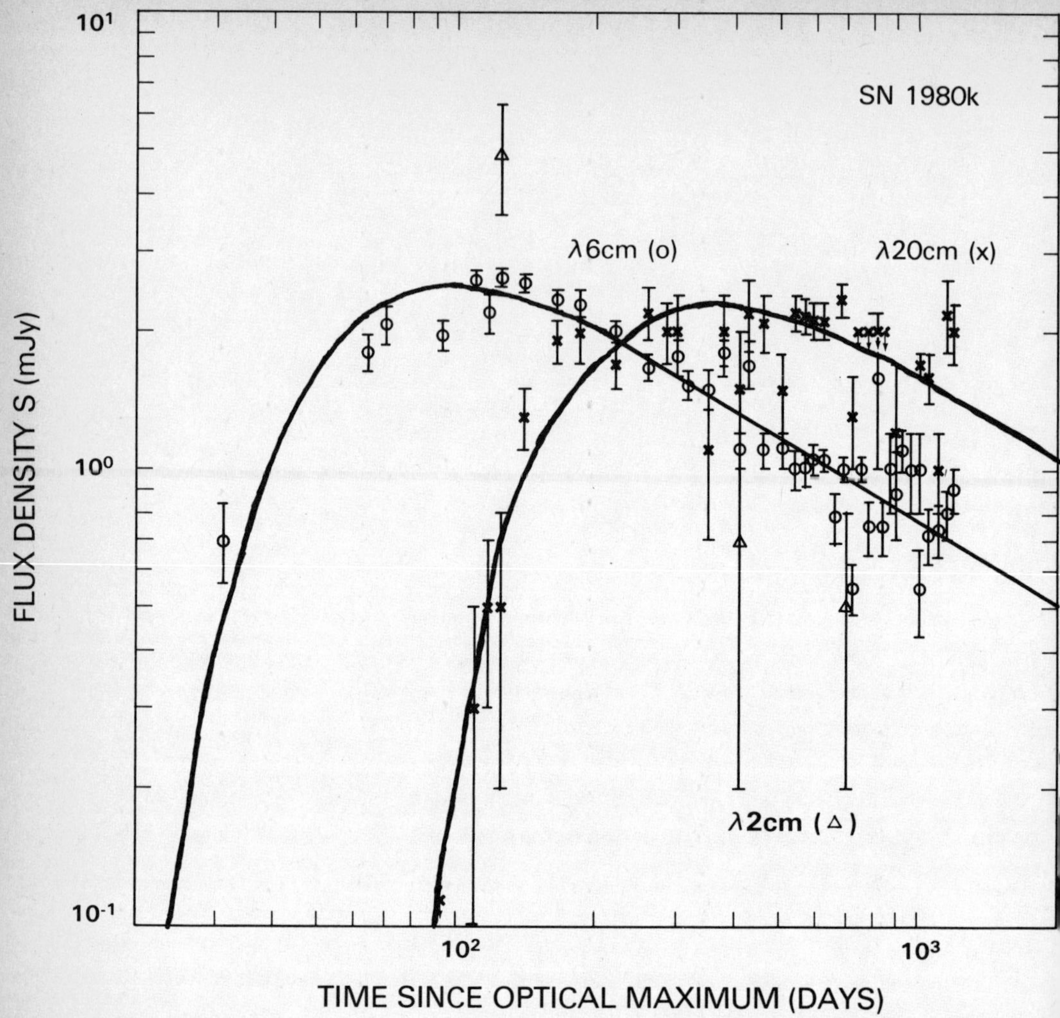

Figure 3. Radio light curves for the Type II supernova SN1980k in NGC6946. All three wavelengths measured with the VLA are shown together as in Fig. 1. The age of the supernova is measured in days from the date of maximum optical light on 1980 November 5. The solid lines represent the "best" fit of curves of the form $S = K_1 \nu^{\alpha} t^{\beta} e^{-\tau}$ where $\tau = K_2 \nu^{-2} t^{\delta}$ with $K_1 = 1.58*10^2$, $\alpha = -0.59$, $\beta = -0.63$, $K_2 = 4.54*10^4$, and $\delta = -1.89$.

extragalactic radio sources and supernova remnants (van der Laan, 1966), and is not necessarily applicable here. In fact, the simplest version of the model predicts lower maximum flux densities at longer wavelengths and a rapid decay after maximum, neither of which is observed.

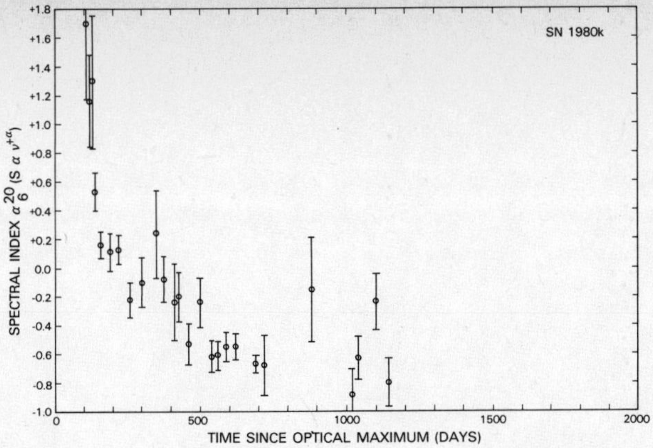

Figure 4: Spectral index (α) for SN1980k between 20 cm and 6 cm plotted as a function of time in days since maximum optical light on 1980 November 5.

A second model is that developed for the Crab Nebula by Pacini and Salvati (1973) which, in its simplest form, considers a balloon of relativistic particles and magnetic fields being blown up by a very rapid pulsar (Pacini and Salvati, 1981). A comparison of the predictions of this model with the observed properties of the radio supernovae shows that it can indeed be made to fit with reasonable accuracy. However, although it cannot be rejected out of hand, it contains one possibly serious flaw. That is, since one expects several solar masses of material to be present in the photosphere and in the immediate vicinity of a Type II supernova, it is difficult to understand how the radio radiation can escape. If even a small fraction of this material around a supernova were ionized, it would keep any centrally-driven remnant from being seen externally at radio wavelengths for decades. Two possibilities have been suggested to solve this problem. One possibility (Bandiera, Pacini, and Salvati, 1983) is that the mass of the supernova clumps quickly into very dense filaments between which the radio radiation can escape. For example, we can look into the center of the Crab Nebula and see the pulsar between the optical filaments even though they are very dense and contain, in total, a great amount of matter. A second suggestion has been made by Shklovskii (1981) which proposes that the radiation itself does not escape from the center of the supernova but that the magnetic field and particles generated by the central pulsar escape to the outside of the supernova photosphere where they generate radio emission which is visible through relatively little matter. Neither of these two models is eliminated by the available data but some workers doubt that either can be effective so quickly in the presence of so much matter.

A third type of model, which avoids the absorbing mass problem by generating the radio emission external to the supernova photosphere, is that by Chevalier

(1981a,b; 1984). In this model the generation of radio radiation takes place in a shock wave in the circumstellar "cocoon" of matter built up by the stellar wind of the supernova progenitor in its last stages of evolution. Since this particle acceleration and magnetic field generation region is outside of the main mass of the supernova, the radiation can be viewed through relatively little matter. It turns out that, for some values of the adjustable parameters describing the properties of this model, the predicted light curves are much the same as those obtained from the centrally-driven model so that it fits the data for Type II supernovae to roughly the same degree of accuracy as the pulsar-driven model.

To distinguish these models, I will refer to them as the "mini-shell" for the shock-wave-driven model, because of its expected shell-like form resembling shell-type supernova remnants such as Cassiopeia A (Cas A), and the "mini-plerion" for the centrally-driven model, because of its expected resemblance to the Crab Nebula and other plerionic supernova remnants. The question, of course, is how well these models fit the data. In Table 1 the parameters are given for both classes of models with, for the mini-shell model, two values of its free parameter m, which is expected to range between 0.75 and 1.0. For the mini-plerion model, one set of values, for the most probable case of the observing frequency ν being less than the break frequency ν_b (See Chevalier, 1984 and Pacini and Salvati, 1973 for more detailed discussion of m and ν_b, respectively), is given. Also given in Table 1 are the actual fitted values for the two Type II radio supernovae SN1980k and SN1979c. As can be seen from the table, the mini-shell model with m = 1 provides a good fit to the observations for 1980k and with m = 0.88 to the observations for SN1979c, while the mini-plerion model provides a reasonable fit to both data sets.

Table 1. Model Predictions and Data Fitting

SN Name	SN Type	α [$S \propto \nu^{+\alpha}$]	γ [$\gamma = 1-2\alpha$]	Mini-Shell [$\beta = -(\gamma+5-6\,m)/2$] m = 1.0		m = 0.88		Mini-Plerion $\nu < \nu_b$	
				α	β	α	β	α	β
Models		-0.6	2.2	-0.6	-0.6	-0.6	-0.92	-0.6	-0.6
Observed	($S \propto \nu^{\alpha} t^{\beta} e^{-\tau}$ for $\tau \rightarrow 0$)								
SN1979c	II	-0.56	2.12			-0.56	-0.92	-0.56	-0.92
SN1980k	II	-0.59	2.18	-0.59	-0.63			-0.59	-0.63

The model curves, assuming that the optically thick turn-on part of the light curves is due to thermal free-free absorption, are shown in Figures 1 and 3 as the solid lines at 6 and 20 cm. A more detailed discussion of the data, models, light curves, etc. is given in a lengthier article in preparation (Weiler et al., 1985).

As can be seen from the curves drawn on Figures 1 and 3, the fits to the data from these two classes of models are reasonable, but certainly not perfect. It is apparent that either the mini-shell or the mini-plerion models can be made to fit the data reasonably well, but that the actual situation in the RSN is considerably more complicated.

This briefly summarizes the status of affairs observationally and theoretically. Now we should turn to a discussion of the use of the radio emission from Type II supernovae as distance indicators, the purpose of the conference. We should look at several questions.

1) Q: Is the turn-on, optically thick part of the radio light curves likely to give us any sort of standard information which might be converted into distance?

A: This does not seem likely. If the turn-on is due to thermal free-free absorption in the material built up by the stellar wind of the progenitor star, as seems likely in all models (see Weiler et al., 1985 for a more detailed discussion of mechanisms), the optically thick data is not directly telling us about the explosion. Also, since Type II supernovae are observed to be quite variable in their optical properties and are probably created by stars with a considerable range of initial stellar masses and stellar winds, they are likely to be quite variable in their individual turn-on times and shapes. In practice, we see a variation in 6 cm turn-on times by more than a factor of 10 from one month for SN1980k to one year for SN1979c. There is, however, a possibility that the turn-on time can give a calibration of the radio luminosity of a Type II supernova and, as a result, give a distance indication, as has been suggested by Chevalier. We will discuss this possibility later.

2) Q: Does some part of the optically thin emission show a consistent set of properties which can help with distance determinations?

A: Here we are hampered by poor statistics, poor data from most known examples, and poor encouragement from theory. For example, the mini-shell models are likely to be quite variable in their intrinsic radio luminosities because of their dependence on the density in the final stellar wind cocoon (see below for the possibility of calibrating such variations from the turn-on times) and thus are not likely to serve as standard candles. The mini-plerion models could, in principle, have reasonably consistent properties if one feels that all pulsars are born alike (which is, of course, disputed), but examination of the observational results is not encouraging. Table 2 shows all of the known or suspected Type II supernovae which have ever been detected in the radio range along with their distances, parent galaxies, ages at the time of detection, flux densities at that age, and spectral luminosities at 6 cm. Also, for ease of comparison, these last are converted into a ratio to the 6 cm spectral luminosity of Cassiopeia A, a well-known supernova remnant. The table shows a wide variation in spectral luminosities and no evidence of Type II RSN being standard candles in the radio range. However, it may quite rightly be argued that these supernovae are of very widely differing ages with SN1950b being more than 30 years old at the time of its first detection while SN1980k was less than 0.2 years old. Thus it is more reasonable to make such a comparison of RSN at similar ages. To do this we have assumed a power law decay of $t^{-0.8}$ and extrapolated all supernova maxima or observations backwards or forwards, as necessary, to an age of one year. These "corrected" spectral luminosities and their ratios to Cas A are given as the last two columns in Table 2. Although the scatter has now been reduced somewhat, it is still too large for Type II RSN to provide anything near a standard candle for distance estimates.

Table 2. Intrinsic Properties of Radio Supernovae

SN Name	SN Type	Dist	Parent Galaxy	6 cm Observed Max.				At Age 1 Year if $S \propto t^{-0.8}$	
				Age	S_6	F_6	Ratio to Cas A	S_6	Ratio to Cas A
		Mpc		years	mJy	erg $s^{-1}Hz^{-1}$		mJy	
SN1950b	II?	~ 7	M83	34	0.5	$3 * 10^{25}$	4	8.4	62
SN1957d	II?	~ 7	M83	27	2	$1 * 10^{26}$	15	28	175
SN1970g	II	7.2	M101	1.4	~ 6	$3 * 10^{26}$	40	7.9	50
SN1979c	II	~ 16	M100	1.2	8.3	$2 * 10^{27}$	250	9.6	290
SN1980k	II	11	N6946	0.2	2.1	$4 * 10^{26}$	45	0.6	14

Besides these difficulties, it must be remembered that most of the RSN are in galaxies which are members of the local group so that the information which they might provide on the Hubble flow is limited. With present instrumental sensitivity, it is difficult even for the VLA to study the details of RSN much beyond 20 Mpc so that we are not optimistic about using the light curves of Type II RSN, or for that matter of any type of RSN, as determinants for H_0.

A possibility which might somewhat improve this situation is the suggestion by Chevalier from his mini-shell models, that the luminosity of the radio emission is dependent on the density in the circumstellar cocoon. Since the turn-on time for such a mini-shell is also dependent on the density in the circumstellar cocoon, the model suggests that a brighter supernova should turn on later and a weaker supernova should turn on earlier. Thus, there is the possibility that the turn-on time at, for example, 6 cm may be related to the intrinsic radio luminosity of the supernova and allow its calibration as a distance indicator. Our present well-studied examples of Type II RSN are consistent with this; SN1979c turned on later than 1980k and was also intrinsically brighter at maximum at 6 cm. However, in order to further investigate this possibility and calibrate it if the relation holds, more extensive observations must be taken.

IV. SUMMARY

There are now available detailed, multi-frequency radio observations on a few radio supernovae and more limited data sets available on several more. Workers in the field have developed sophisticated models to explain the observed phenomena, but an observational distinction between the two main classes of centrally- ("pulsar") driven mini-plerion and externally- (shock) driven mini-shell models is not yet possible. Neither theoretical nor observational results provide much encouragement for using the radio properties of Type II supernovae as distance indicators. However, the possibility of using turn-on times for calibrating individual RSN luminosities for distance determination still remains and cannot yet be proven or disproven from the limited data now available.

At present, the most promising method for obtaining distance information from RSN appears to be that of measuring angular sizes and velocities by VLBI techniques as discusssed elsewhere in this volume by Bartel and in Bartel et al. (1983).

ACKNOWLEDGEMENTS

I wish to thank my collaborators, Dr. R.A. Sramek of the National Radio Astronomy Observatory (NRAO), Dr. J.M. van der Hulst of the Netherlands Foundation for Radio Astronomy (SRZM), and Dr. N. Panagia of the Istituto di Radioastronomia in Bologna, Italy for their active and continuing help and cooperation in taking, reducing, analyzing, and interpreting the data. Also, I am deeply indebted to Ms. P. Goheen of the National Science Foundation for her rapid and accurate preparation of the manuscript.

----

[1]The VLA is operated by the National Radio Astronomy Observatory through the Associated Universities, Inc. under contract to the National Science Foundation.

REFERENCES

Allen, R.J., Goss, W.M., Ekers, R.D., and de Bruyn, A.G. 1976, Astr. Ap. 48, 253.

Bandiera, R., Pacini, F., and Salvati, M. 1983, Astr. Ap. 126, 7.

Bartel, N., Gorenstein, M.V., Marcaide, J.M., Rogers, A.E.E., Shapiro, I.I., and Weiler, K.W. 1983, Bull. Am. Astr. Soc. 15, 954.

Brown, R.L. and Marscher, A.P. 1978, Ap.J. 220, 467.

Chevalier, R.A. 1981a, Ap.J. 251, 259.

Chevalier, R.A. 1981b, Ap.J. 246, 267.

Chevalier, R.A. 1984, Ap.J. (Letters), 285, L63.

de Bruyn, A.G. 1973, Astr. Ap. 26, 105.

Gottesman, S.T., Broderick, J.J., Brown, R.L., Balick, B., and Palmer, P. 1972, Ap.J. 174, 383.

Pacini, F. and Salvati, M. 1973, Ap.J. 186, 249.

Pacini, F. and Salvati, M. 1981, Ap.J. (Letters) 245, L107.

Shklovskii, I.S. 1981, Sov. Astr. Lett. 7(4), 263.

Ulmer, M.P., Crane, P.C., Brown, R.L., and van der Hulst, J.M. 1980, Nature 285, 151.

van der Laan, H. 1966, Nature 211, 1131.

Weiler, K.W., Sramek, R.A., Panagia, N., and van der Hulst, J.M. 1985, in preparation.

RADIO OBSERVATIONS OF HISTORICAL, EXTRAGALACTIC SUPERNOVAE

John J. Cowan and David Branch
Department of Physics and Astronomy
University of Oklahoma

ABSTRACT

VLA maps of the galaxy M83 (NGC 5236) at 6 and 20 cm reveal the presence of both non-thermal and thermal sources, lying predominantly along the inner edges of the optical spiral arms. A radio source coincident with the optical position of supernova 1957d is found to have a non-thermal spectrum; thus SN 1957d is confirmed as the first supernova of intermediate-age (~ 10-300 years) to be detected at any wavelength. A second non-thermal source is tentatively identified with supernova 1950b, pending the measurement of a precise optical position of the supernova. Two other non-thermal sources lying along the inner edges of the spiral arms are likely to be the remnants of supernovae which were not seen optically. All four sources have monochromatic luminosities comparable to that of Cas A. A composite "radio light curve" for supernovae and young supernova remnants of known age is presented and discussed.

I. INTRODUCTION

The study of radio emission from supernovae of known intermediate age (~10-300 years) is in its infancy. Cas A, at an age of $\geq$ 300 years (van den Bergh and Kamper 1983), is the youngest remnant of known age in the Galaxy. Among the four extragalactic "radio supernovae" which have been detected shortly after the time of maximum optical light (Gottesman et al. 1972; Weiler et al. 1981; 1982; Sramek, Panagia, and Weiler 1984), SN 1979c in M100, at its present age of six years, has been followed for the longest period of time. Until recently radio emission from supernovae of intermediate age had not been detected, in spite of many attempts (e.g., De Bruyn 1973; Brown and Marscher 1978; Ulmer et al. 1980).

In 1981 we used the Very Large Array[1] (VLA) to search for emission at 20 cm from the four optical supernovae which had been seen between 1923 and 1968 in the relatively nearby (~ 4-8 Mpc) Centaurus-group galaxy M83 (NGC 5236), and from the two supernovae (1895b and 1972e) in its companion galaxy NGC 5253. Two weak sources were found in the direction of M83, but neither coincided with the published positions of its historical supernovae (Cowan and Branch 1982). However, Pennington and Dufour (1983) measured the position of SN 1957d in M83 from archival plates and found that the published position was in error by more than a minute of arc. The new, precise posi-

[1]The National Radio Astronomy Observatory is operated by Associated Universities, Inc. under contract with the National Science Foundation.

tion measured by Pennington and Dufour lies only three arc seconds from one of the radio sources we detected in M83. Within the uncertainties of the radio and the optical positions, this radio source is coincident with the measured position of the supernova. Thus, SN 1957d is the first supernova of known intermediate age to be detected at any wavelength.

The discovery of radio emission from a supernova 24 years old, at a monochromatic luminosity comparable to that of Cas A, has raised hopes that additional historical supernovae are within the range of the VLA, and that the entire radio evolution of supernovae can be determined observationally. This paper reports a first step in that direction. We have used the VLA to make further observations of M83, at 6 and 20 cm, to determine a spectral index for the source associated with SN 1957d and to search deeper for radio emission from the other historical supernovae.

M83 is an SAB(s)c I-II galaxy which appears to have a high rate of star formation (Talbot, Jensen, and Dufour 1979; Condon 1983) and an active or "star-burst" nucleus (Pastoriza 1975; Rieke 1976; Condon et al. 1982; Bohlin et al. 1984). This apparent high rate of star formation may be responsible for the many supernovae seen in this galaxy. In July, 1983, a fifth optical supernova, SN 1983n, was discovered in M83 by R. Evans (Thompson 1983). Radio emission from this (optically peculiar) Type I supernova has been reported by Sramek et al. (1984) and interpreted by Chevalier (1984a). Although SN 1983n has been detected in our recent 6- and 20-cm observations, the emphasis in this paper is on the older, intermediate supernovae.

The observations are described in § II. In § III, the results of the observations are displayed. An attempt is made to associate the observed non-thermal radio sources with supernovae. A summary of our current knowledge of the radio behavior of supernovae of intermediate age is given in § IV.

II. OBSERVATIONS

We first observed M83 on 1981 April 29 for three hours (Cowan and Branch 1982). The array was in the hybrid A/B configuration with 22 antennas in operation. The hybrid configuration with the extended northern arm is well-suited for sources such as M83 with low declinations. The observations were made at a frequency of 1.5 GHz (20 cm) using a 12.5 MHz band width. The rms noise level near the center of the field was approximately 0.3 mJy. An additional six hours of observations at 20 cm were made on 1983 December 15. The array was again in the A/B configuration and 27 antennas were used. Owing to improvements in the system and the longer time of observation the rms noise level was reduced to less than 0.1 mJy. Finally, six and one-half hours of observations at 6 cm were made on 1984 March 13, when the array was in the hybrid B/C configuration. A band width of 25 MHz and 26 antennas were used, and the rms noise level was again less than 0.1 mJy.

The phase calibration source used in the observations was 1354-132. 3C 286 was used to establish the flux density scale with assumed fluxes of 7.4 and 14.5 Jy at 6 and 20 cm respectively. At 6 cm in the B/C array, the synthesized beam size was 3.9" x 2.8". The radio positions listed in this paper are based upon the calibrator source, 1354-132, which should be known to an accuracy of 0.1". The error in the positional accuracy of the sources, however, can be estimated to be approximately one-fourth of the synthesized beam size or as much as 1".

III. RESULTS

A 6-cm contour map of the M83 field is shown in Figure 1. Nonthermal sources are denoted by Greek letters and thermal sources by numbers. The dashed lines indicate the approximate positions of the inner edges of the spiral arms of M83. The relationship between the radio sources and the optical characteristics of the galaxy can be seen more directly in Figure 2 (Plate xx), where the 6-cm radio contour map has been placed over the optical image of M83 from the Hubble Atlas of Galaxies (Sandage 1961). Note that most of the sources (both thermal and non-thermal) lie along the inner edges of the spiral arms. The radio properties of the nonthermal sources are listed in Table 1. The radio fluxes listed in the tables all have associated errors of ± 0.1 mJy.

a) Nonthermal Sources

It is always possible that one or more of the non-thermal sources are background sources. From the deep VLA survey of Condon and Mitchell (1984) we estimate that in an area of 25 square arc-minutes (the approximate size of the optical image of M83) the expected number of background sources brighter than 0.6 mJy is 0.5. However, four of the five non-thermal sources listed in Table 1 appear on the inner edges of the spiral arms of M83, and the fifth (ε) is unquestionably SN 1983n. In the following discussion we therefore assume that all of the non-thermal sources are associated with M83. At a distance between 4 and 8 Mpc their monochromatic luminosities are comparable to that of Cas A.

The five optical supernovae which have been seen in M83 are listed in Table 2. The optical positions for SN 1923a, 1957d, and 1983n are astrometric positions of arc-second accuracy. The positions of SN 1950b and 1968l have been published only as offsets with respect to the optical center of M83 (Kowal 1984). The coordinates given in Table 2 for SNe 1950b and 1968l are based on an adopted position of the optical center of $13^h34^m11^s.55$, -29°36'42.2". The positions of these two supernovae are only of about 10 arc-second accuracy because the optical center of M83 is not well defined (see § III-B).

i) α

The unresolved source α was detected in our original 20-cm observations (Cowan and Branch 1982). It was not recognized to be associated with a historical supernova due to the incorrect published position of SN 1957d. Pennington and Dufour's (1983) new position put SN 1957d 3 arc-sec-

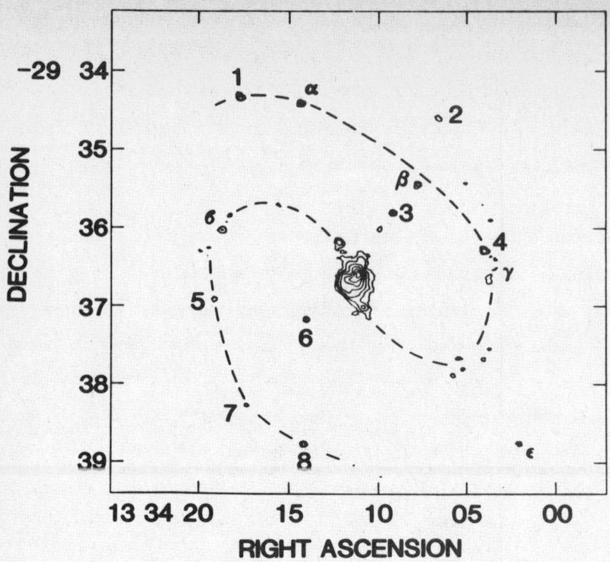

Figure 1. Observations of M83 at 6 cm. Nonthermal sources are indicated by Greek letters, thermal sources by numbers. The dashed lines represent the approximate position of the inner edges of the spiral arms. Coordinates are 1950 epoch.

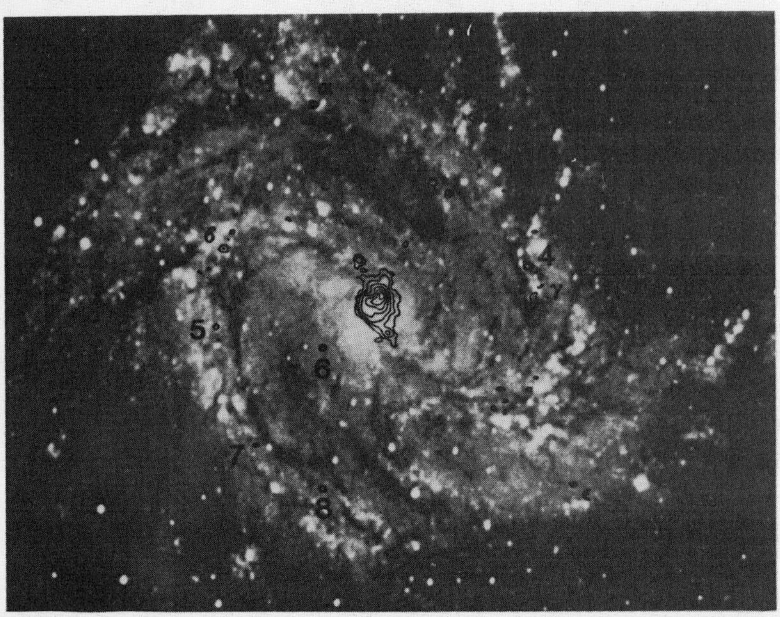

Figure 2. Overlay of the optical image and the 6 cm radio map of M83. Nonthermal point sources are indicated by Greek letters, thermal sources by numbers.

TABLE 1

NONTHERMAL SOURCES IN M83

Source	Radio Coordinates Right Ascension			Declination			Flux Density (mJy)[a] 20 cm	6 cm	Spectral Index ($S \propto \nu^{+\alpha}$)
α(SN 1957d)	13^h	34^m	14.3^s	-29°	34'	24.0"	2.6	1.9	-0.3
β	13	34	07.8	-29	35	27.0	3.1	0.8	-1.1
γ(SN1950b)	13	34	03.8	-29	36	40.5	0.8	0.5	-0.4
δ(near SN 1923a)	13	34	18.7	-29	36	00.0	0.9	0.7	-0.3
ε(SN 1983n)	13	34	02.0	-29	38	45.0	4.2	0.6	-1.0

[a]±0.1 mJy

TABLE 2

OPTICAL SUPERNOVAE IN M83

SN	1950 Coordinates Right Ascension			Declination			Type
1923a[1]	13^h	34^m	20.0^s	-29°	35'	48.0"	II
1950b	13	34	03.5	-29	36	42.2	-
1957d[2]	13	34	14.4	-29	34	27.0	-
1968ℓ	13	34	11.2	-29	36	42.2	II
1983n[3]	13	34	01.7	-29	38	48.0	I

[1]Pennington, Talbot and Dufour (1982)

[2]Pennington and Dufour (1983)

[3]Cragg (1983)

onds from α, strongly suggesting an identification, but no spectral index was available at that time. Our recent 6 cm measurements allow the determination of the spectral index, -0.25, and verify the nonthermal nature of the source. The identification of the radio source α with the optical supernova 1957d is now confirmed. For a discussion of the optical characteristics of the region in which SN 1957d occurred, see Pennington and Dufour (1983).

ii) β

The source β also was detected in the original 20-cm observations. Its spectral index now is found to be -1.07. No optical supernova has been reported near the position of β. Nevertheless, its position along the inner edge of a spiral arm and its non-thermal spectrum suggest that β is likely to be a supernova remnant in M83 (see also Pennington and Dufour 1983). A 20-cm contour map of this source (Figure 3) shows that it has a jet or tail associated with it. The tail may be physically associated with the supernova itself, or the source may be superimposed on an H II region.

iii) γ

The source γ is weaker than α and β and was not detected in the original 20 cm observations. Its spectral index is -0.38 . This non-thermal point source seems to be superimposed on a more extended region of emission, perhaps from an H II region. The position of β differs from that of SN 1950b by less than 5 arc-seconds. Since the position of SN 1950b is derived from offsets from the center of M83 and is not accurately known, we tentatively regard SN 1950b to be the second intermediate-age supernova to be detected at radio wavelenths. An astrometric position for SN 1950b from archival plates is needed to establish positively the identification.

iv) δ

The source δ is another relatively weak source not seen in the original 20-cm observations. Its spectral index of -0.27 indicates a non-thermal but relatively flat spectrum, like that of α (SN 1957d) and γ (SN 1950b). LIke γ, δ appears to be superimposed on an HII region. The position of δ is only 21 arc-seconds from that of SN 1923a, but Pennington et al. (1982) quote an accuracy of 0.25 arc-seconds for their astrometric position of SN 1923a. γ is likely to be the remnant of a supernova which was not seen optically.

v) ε

This unresolved source corresponds to the recent supernova 1983n, the first Type I radio supernova to be detected (Sramek et al. 1984). Figure 1 and Figure 2 (Plate xx) show that unlike the other non-thermal sources this Type I radio supernova does not lie along the inner edge of a spiral arm. It does, however, lie within an arm. (The outlying spiral arms of M83 are well delineated by the Hα map published by de Vaucouleurs, Pence, and Davoust 1983).

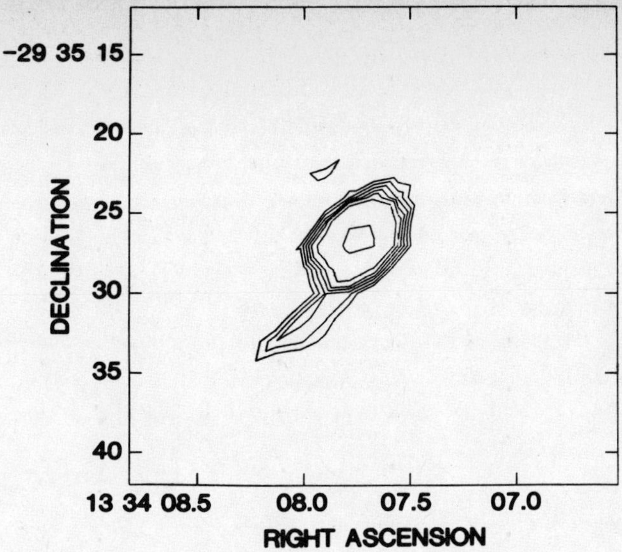

Figure 3. Radio contours at 20 cm of the nonthermal point source β. Coordinates are 1950 epoch.

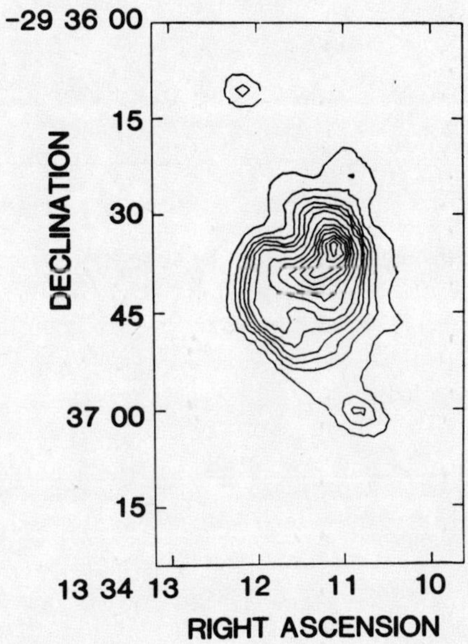

Figure 4. Radio contours at 6 cm of the central regions of M83. The peak flux is 1.49 x 10^{-2} Jy per beam. Coordinates are 1950 epoch.

B. The Nucleus

A 6-cm contour map of the central regions of M83 is shown in Figure 4. The location of the peak radio emission is $\alpha=13^h34^m11.1^s$, $\delta=-29°36'34.5''$, in agreement with that of Condon et al. (1982) at 6 cm, and with our 20-cm position. Cowan and Branch (1982) noted that the radio and optical positions of the center of M83 differ. Recent determinations of the center of M83 are listed in Table 3. The radio position differs from the optical position reported by Gallouet et al. (1975) by 12 arc-seconds in right ascension and more than 14 arc-seconds in declination, while the reported error limits are only 6 seconds in right ascension and 5 seconds in declination. The center determined from the Hα surface photometry of Talbot et al. (1979) is $13^h34^m11^s.55$, $-29°36'42.2''$ (Rumstay and Kaufman 1983). This is closer to the radio position, but still different by 7 arc-seconds in both coordinates. The optical nucleus of M83 is quite complex (see Wood and Andrews 1974), and the positions of peak optical and radio emission may really be offset.

The peak radio flux from the nucleus was 2.72×10^{-2} Jy per beam and 1.49×10^{-2} Jy per beam at 20 and 6 cm respectively, and the spectral index of the nucleus is -0.5.

TABLE 3

RADIO AND OPTICAL POSITIONS OF THE CENTER OF M83

Feature	**1950 Coordinates**					
	Right Ascension			**Declination**		
Optical Center[1]	13^h	34^m	10.2^s	$-29°$	36'	49.0"
Optical Center[2]	13	34	11.6	-29	36	42.2
Radio Center[3]	13	34	11.1	-29	36	35.2
Radio Center[4]	13	34	11.1	-29	36	34.5

[1]Gallouet et al. (1975): Blue Palomar Observatory Sky Survey Plates

[2]Talbot et al. (1979) (See Rumstay and Kaufman 1983): Hα Surface Photometry

[3]Condon et al. (1982): 6 cm

[4]This paper: 6 cm and 20 cm

IV. DISCUSSION AND CONCLUSIONS

Type I supernovae (SNe I) are found in all kinds of galaxies. Those in spirals show no tendency to concentrate in the arms (Maza and van den Bergh 1976). The peculiar Type I supernova 1983n (source ε) is located in a spiral arm, but not along the inner edge. SNe II almost always appear in the arms, and therefore are commonly attributed to the explosions of massive stars ($\geq 8\ M_{\odot}$). The two historical supernovae in M83 which have not been detected at radio wavelengths, SN 1923a and 1968l, both are classified as Type II (Minkowski 1964; Wood and Andrews 1974). The appearance of SN 1968l near the nucleus of M83, rather than on an arm, was exceptional among SNe II. The two supernovae which have been detected, SN 1957d (source α) and 1950b (source γ), were not well observed optically, and cannot be classified. Considering that SN 1957d and 1950b as well as the optically unidentified non-thermal sources β and δ all occurred on the inner edges of spiral arms, and that the radio emission of the Type I supernova 1983n is rapidly fading, it is possible that radio bright intermediate-age supernovae all derive from the explosions of massive stars. It is clear, in any case, that the radio luminosities of intermediate-age supernovae cover a significant range in luminosity. It will be interesting and informative to compare the optical and radio properties of supernovae when a larger sample of historical supernovae has been detected at radio wavelengths.

Understanding the evolution of these historical supernovae may also help in establishing the distance scale of the universe. The time dependence of the radio emission of the supernovae will provide insight into the nature of the interaction between the supernova shock and the surrounding material, and thus should help to determine such physical parameters as the actual expansion velocity of the supernova shock wave. Since VLBI observations can determine the change in angular size of some these sources, the distance to these supernovae and hence their parent galaxies could then be accurately determined. In this way, new determinations of the Hubble constant may be obtained.

In Figure 5, we attempt to summarize our current fragmentary knowledge of the radio evolution of supernovae by plotting a "radio light curve" for supernovae and young supernova remnants of known age. The distances of the Tycho and Kepler remnants are taken to be 3 kpc, and the distances of the four other Galactic remnants (Cas A, the Crab, and the remnants of SNe 1181 and 1006) are from Milne (1979). The luminous young extragalactic remnant in NGC 4449 is also included, at an assumed distance of 5 Mpc; strictly, its age is not known because the supernova itself was not observed, but the age is constrained to be near 120 years (Blair, Kirshner, and Winkler 1983). The upper limit for SN 1885a in the Andromeda Galaxy is from Dickel and d'Odorico (1984), with their 6-cm upper limit being transformed to 20-cm by assuming a spectral index of - 1. SN 1957d and 1950b in M83 are shown as detections, while the other historical supernovae in M83 and NGC 5253 (Cowan and Branch 1982) are shown as upper limits. The distances to M83 and NGC 6946 are taken to be 4 and 8 Mpc respectively; distances to other galaxies beyond the Local Group are based on a Hubble constant of 60 km s^{-1} Mpc^{-1}.

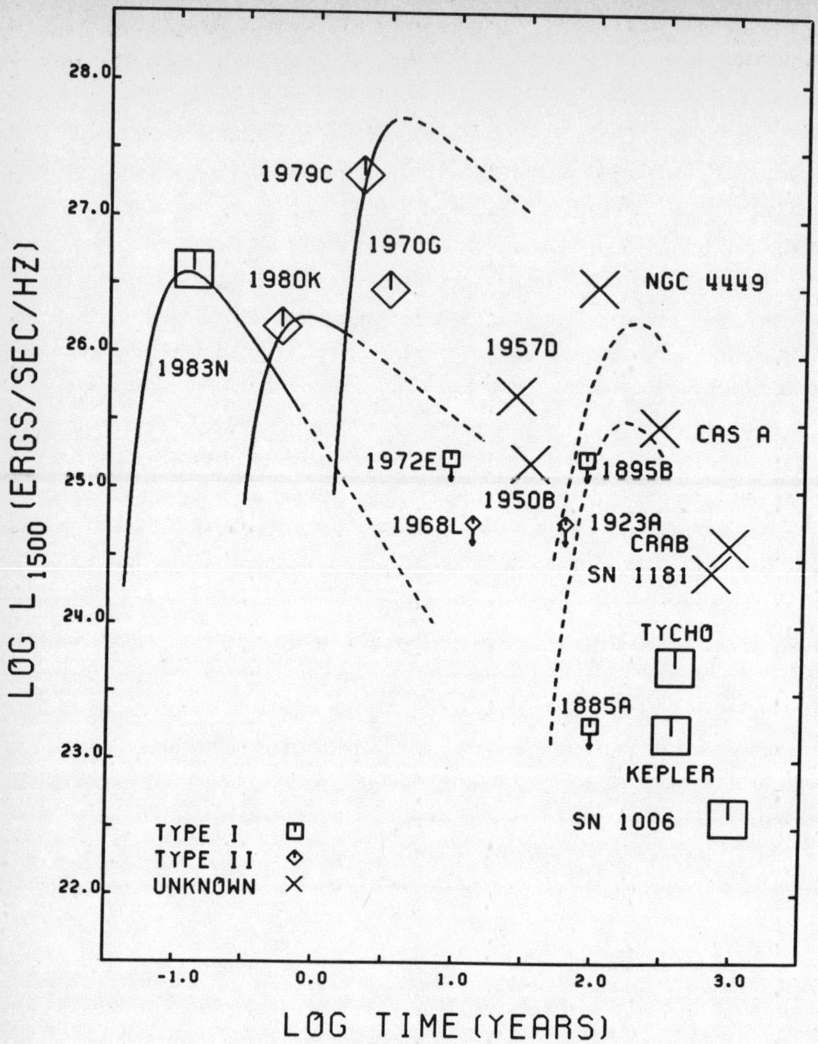

Figure 5. Radio light curves of supernovae and supernova remnants. Monochromatic (20 cm) absolute luminosity is plotted against time since optical maximum. Upper limits are denoted by small symbols with arrows. All other points indicate detections. The 3 solid curves are fits to the observations of supernovae using the theoretical model of Chevalier (1984). The dashed extensions to the three curves are extrapolations of the theory. The two dashed curves at the right are theoretical predictions for the time evolution of the radio emission of young supernova remnants, after Cowsik and Sarkar (1984).

For the four radio supernovae which have been detected near maximum optical light (SNe 1970g in M101, 1979c in M100, 1980k in NGC 6946, and 1983n in M83), the symbol is plotted at the position of the peak radio luminosity. For SNe 1979c, 1980k, and 1983n the associated curve is computed from Chevalier's (1984b),expression for the flux density, based on his model for the radio emission as synchrotron radiation coming from the hot region of interaction between the ejected supernova shell and a surrounding circumstellar shell that resulted from pre-supernova mass loss. The solid portion of the curve is the part that has been observed; the theory gives a good representation of the observations. The dashed parts of the curves are extrapolations based on the theory, using the parameters chosen by Chevalier (1984a,b). We see that the extrapolated radio light curves for SNe 1979c and 1980k, both of Type II, bracket the positions of SN 1957d and 1950b. These may be two fading radio supernovae.

There are various theories that attempt to explain the radio emission from young supernova remnants (e.g. Pacini and Salvati 1973; Reynolds and Chevalier 1984). Figure 5 shows the evolution of the 20-cm luminosity of models (c) and (d) of Cowsik and Sarkar (1984). These models, indicated by dashed curves, bracket Cas A. Following Gull (1973), the Cowsik and Sarkar models turn on as synchrotron-radiation radio sources when the mass of swept-up interstellar material exceeds the mass of ejected material. Rayleigh-Taylor instabilities in the decelerating ejecta transform about 1 per cent of the explosion energy into small-scale turbulent motions, and magnetic-field amplification and the generation of relativistic electrons proceeds until equipartition of turbulent, magnetic, and particle energies is achieved. See Cowsik and Sarkar (1984) for the assumed rates of particle injection and acceleration on which the models are based. Their theoretical curves shown in Figure 5 do not predict that remnants of supernovae as young as SNe 1957d and 1950b should have already turned on as radio sources, but the theoretical peak luminosity and the age at which it occurs depend on the explosion energy, the ejected mass, and the density of the surrounding interstellar medium. The peak luminosity scales as

$$L_\nu(\max) \propto E^{(1-5\alpha)/2} M^{-(1-\alpha)/2} \rho^{(1-\alpha)/2}$$

and the age at which the peak is reached is given by

$$t(\max) = 180\ E_{51}{}^{-1/2} M^{5/6} \rho^{-1/3}\ \text{years}$$

where E_{51} is the explosion energy in units of 10^{51} ergs, M is the ejected mass in solar masses, ρ is the interstellar density in units of 1 baryon cm^{-3}, and the spectral index $\alpha \sim -1$. The Cowsik and Sarkar models shown in Figure 5 are for 10^{51} ergs, 0.5 solar masses, and an interstellar density of 1 baryon cm^{-3}. The theoretical peak could be moved to an age significantly less than 100 years by invoking some combination of a higher explosion energy, a smaller ejected mass, and a higher density of the interstellar medium.

At present, radio emission from a supernova of intermediate age could be attributed plausibly to either the slow fading of a radio supernova, in which the interaction is between the ejected matter and circumstellar material, or as the turning on of a young supernova remnant, in which the ejected matter is interacting with the interstellar medium. Observations of the time dependence of the radio luminosities of SNe 1957d and 1950b will be needed to choose between the two possibilities.

We thank Rick Perley and Dick Sramek for invaluable help with the data reduction and K.L. Venkatakrishna for his assistance. This research was supported in part by NSF grants AST 82-14964 and AST 82-18625.

REFERENCES

Blair, W., Kirshner, R.P., and Winkler, P.F. 1983, Ap.J., **272**, 84.

Bohlin, R.C., Cornett, R.H., Hill, J.K., Smith, A.M. and Stecker, T.P., 1983, Ap.J.(Letters), **274**, L53.

Brown, R.L. and Marscher, A.P. 1978, Ap.J., **220**, 467.

Chevalier, R.A. 1984a, Ap.J.(Letters), **285**, L63.

Chevalier, R.A. 1984b, Ann. N.Y. Acad. Sci, **422**, 215.

Condon, J.J. 1983, Ap.J. Suppl., **53**, 459.

Condon, J.J., Condon, M.A., Gisler, G., and Puschell, J.J. 1982, Ap.J. **252**, 102.

Condon, J.J., and Mitchell, K.J. 1984, preprint.

Cowan, J.J. and Branch, D., 1982, Ap.J., **258**, 31.

Cowsik, K.R. and Sarkar, S. 1984, M.N.R.A.S., **207**, 745.

Cragg, T. 1983, IAU Circ., No. 3835.

de Bruyn, A.G. 1973, Astr. Ap., **26**, 105.

de Vaucouleurs, D.G., Pence, W.D., and Davoust, E., 1983 Ap.J. Suppl., **53**, 17.

Dickel, J.R. and D'Odorico, S., 1984,M.N.R.A.S., **206**, 351.

Gallouet, L., Heidmann, N., and Dampierre, F., 1975, Astron. Ap. Suppl., **19**, 1.

Gottesman, S.T., Broderick, J.J., Brown, R.L., Balick, B., and Palmer, P. 1972, Ap.J., **174**, 383.

Gull, S.F. 1973, M.N.R.A.S., **161**, 47.

Kowal, C.T. 1984, Palomar Supernova List.

Maza, J., and van den Bergh, S. 1976, Ap.J., **204**, 519.

Milne, D.K. 1979, Australian J. Phys.,**32**, 83.

Minkowski, R. 1964, Ann.Rev.Astr.Ap., **2**, 247.

Pacini, F. and Salvati, M., 1973, Ap.J., **186**, 249.

Pastoriza, M.G. 1975, Ap.Space Sci., **33**, 173.

Pennington, R.L. and Dufour, R., 1983, Ap.J.(Letters), **270**, L7.

Pennington, R.L., Talbot, R.J.,Jr, and Dufour, R., 1982, A.J., **87**, 1538.

Reynolds, S.P., and Chevalier, R.A. 1984, Ap.J., **278**, 630.

Rieke, G.H. 1976, Ap.J.(Letters), **206**, L15.

Rumstay, K.S. and Kaufman, M., 1983, Ap.J., **274**, 611.

Sandage, A. 1961, The Hubble Atlas of Galaxies (Washington:Carnegie).

Sramek, R.A., Panagia, N., and Weiler, K.W., 1984, Ap.J.(Letters), **285**, L59.

Talbot, R.J., Jr., Jensen, E.B., and Dufour, R., 1979, Ap.J., **229**, 91.

Thompson, G.D. 1983, IAU Circ., No. 3835.

Ulmer, M.P., Crane, P.C., Brown, R.L., and van der Hulst, J.M. 1980, Nature, **285**, 151.

van den Bergh, S., and Kamper, K.W. 1983, Ap.J., **268**, 129.

Weiler, K.W., van der Hulst, J.M., Sramek, R.A., and Panagia, N. 1981, Ap.J.(Letters), **243**, L151.

Weiler, K.W., Sramek, R.A., van der Hulst, J.M., and Panagia, N. 1982, in Supernovae: A Survey of Current Research, eds. M.J. Rees and R.J. Stoneham (Dordrecht:Reidel), p. 281.

Wood, R., and Andrews, P.J. 1974, M.N.R.A.S., **167**, 13.

DISCOVERY OF AN ENTIRE POPULATION OF RADIO SUPERNOVA CANDIDATES IN THE NUCLEUS OF MESSIER 82

by

Philipp P. Kronberg
Department of Astronomy and Scarborough College
University of Toronto

SUMMARY

Observations with the full resolution of the NRAO Very Large Array in New Mexico by Kronberg, Biermann and Schwab have revealed about 40 discrete radio sources in the inner, visually obscured nucleus of the enigmatic galaxy Messier 82. These radio sources, more luminous than any comparable objects in our Galaxy, are comparable to those of the half dozen radio supernovae found to date in other nearby galaxies. Their physical characteristics are presumably similar to those in more distant galaxies which might be used as distance scale indicators.

Repeated monitoring of the M82 nuclear region since 1981 by R.A. Sramek and myself reveals that most of the brightest of these sources are declining in intensity on the scale of months to years. Both their radio luminosity and variability strongly suggest that M82 contains an entire population of radio supernovae, whose radio structure and variability we can compare at the same distance. Striking individual differences in variability between 1981 and 1983 already suggest that M82's population of radio supernovae contains both Type I and Type II supernovae, and possibly other classes as yet unrecognized.

Most of the other 10 strongest sources are declining so rapidly that they will fade into the background within $\sim$35 years. Thus, we expect new supernovae to appear in M82's nucleus every few years. This will provide the opportunity for VLBI size and structure evolution measurements over the next few years. The high degree of asymmetry seen for many of the M82 sources suggests that the radio size growth of this class of source (presumably radio SN) may not be usable as a distance scale indicator until we acquire a better understanding of their physical nature.

I. INTRODUCTION

Full resolution VLA maps of the irregular galaxy M82 reveal more than 40 discrete radio sources, most of which are more luminous than any radio source in our Galaxy (Kronberg and Biermann, 1983, Kronberg, Biermann and Schwab, 1985(KBS2)). A radio photograph of the 4.88 GHz map made in February 1981 with a resolution of 0.3 arcseconds is shown in Fig. 1. None of the myriad of radio sources has been optically identified, because of the high visual extinction in M82's nucleus. However, the similarity of their radio luminosities to those of recently discovered extragalactic radio supernovae (cf. Weiler, Sramek, van der Hulst and Panagia, 1983) strongly suggests that we are seeing, for the first time, an entire population of radio sources arising from supernovae associated with an intense burst of massive star formation in M82's nucleus.

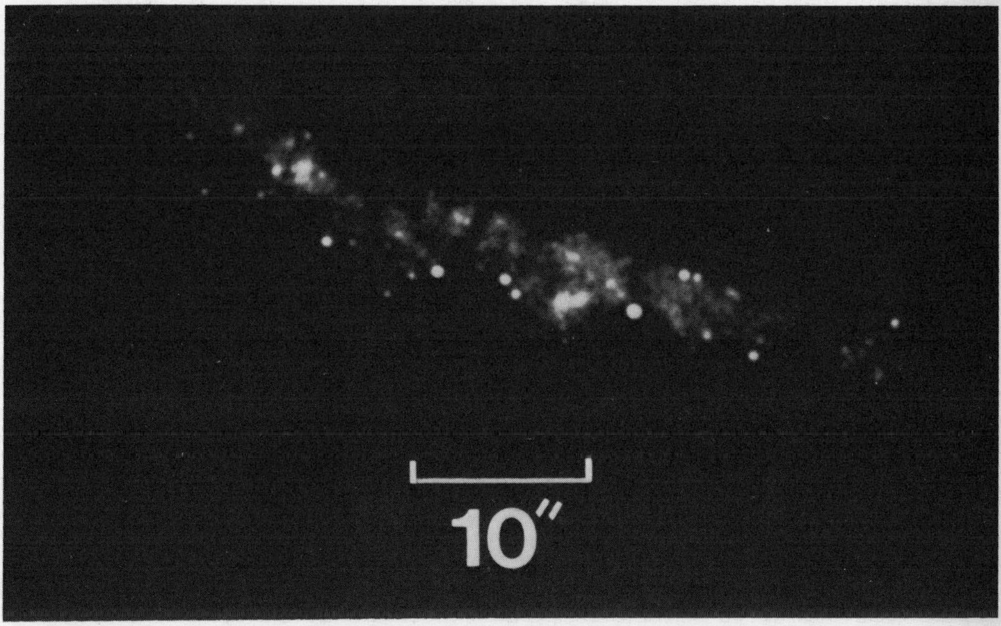

Figure 1. 5GHz VLA map of the nucleus of M82 at epoch 1981.1.
The resolution is 0.34", and at the adopted distance of 3.2 Mpc, 1" = 15 pc.
The faintest discrete sources visible in this picture have flux densities comparable to Cassiopeia A if it were at the distance of M82.

II. OBSERVATIONS

It is well established that the brightest of the M82 nuclear sources, 41.9+58, has been declining in flux density since at least the early 1970's (Kronberg and Clarke, 1978, Kronberg, Biermann and Schwab, 1981(KBS1)). In doing so, it has maintained a nearly constant spectral index (∿0.9) in the optically thin part of its spectrum above 1 GHz.

Since the Spring of 1982, R.A. Sramek and I have done repeated, multi-frequency mapping of M82 with the NRAO Very Large Array (VLA) in order to search for time variability in this large, concentrated population of presumed radio supernovae and supernova remnants. M82 was mapped at 4.88 GHz (6 cm) in 1982 April, May and June, then again in 1983 August and October. Maps have also been made at 1.4 GHz at each of the above epochs, and in 1983 and 1984 repeat observations were made at 15 and 23 GHz. This paper reports some of the preliminary results of the monitoring experiment at λ6cm, which cover a timespan of 2.7 years at the same resolution and sensitivity. Preliminary results on the angular sizes are also discussed, and some comments are made on the potential usefulness of M82's radio supernovae for using similar objects as distance indicators.

III. FLUX DENSITY DETERMINATIONS AND CALIBRATION

The flux density scale of all observations is relative to 3C286. Its adopted integrated (zero-spacing) flux was 7.41 Jy at 4.885 GHz. Each VLA observing run between 1982 April and 1983 October was made within the same 8-hr local sidereal time range, thus giving similar interferometer baseline coverage. Local amplitude and phase calibration in 1981 February was performed via a nearby calibrator 0917+624 (cf. KBS2) whose flux density was tied to 3C286. In 1982 April and succeeding epochs, the local calibrator was 1044+719.

After applying the external calibrations, self-calibration in phase was used to improve the dynamic range of the maps. Interferometer baselines less than 50,000λ were removed to reduce the effect of large structure in the maps. The final maps were then analysed using NRAO-AIPS routines to obtain best-fit integrated and peak flux density values for all sources which were clearly detectable above the ambient background level and which were either smaller than, or comparable to our 0.34" HPBW. (This is 5.3 pc at M82 for an adopted distance of 3.2 Mpc.) In deriving the flux densities, a best fit mean and slope of the local background radiation was subtracted from each source flux. In the following analysis we use peak flux densities since these are less contaminated by the background radiation. Since our sources are unresolved or nearly so, the peak and integrated fluxes are close; in fact we use their ratio as a test for significant extended emission.

For the purpose of intercomparing flux densities at different epochs, there are three sources of error; (1) the uncertainty in the flux scaling, (2) the error of fit to the best peak flux, and (3) the offset error involved in estimating the true background level which was subtracted. The derived flux densities for our calibrators are given in Table I. One means of estimating (1) is to compare the flux densities for several epochs of the secondary calibrator, 1044+719, as determined from 3C286. This can be considered as an upper limit to the calibration scale error, since some of this variation may be intrinsic to the source. The rms scatter from the data listed in Table I below is ±2%, and we have used this value for error component (1).

The combined error due to components (2) and (3) was obtained from the statistics of flux density differences between the closely spaced epochs over which genuine flux density variations are assumed to be insignificant. In our preliminary analysis, we find the 1σ repeatability error for the <u>faint</u> sources to be $\sim$0.2 mJy. It is hoped that component (3) can be reduced further in future if yet higher quality images with greater dynamic range can be obtained.

TABLE I

CALIBRATOR FLUX DENSITIES RELATIVE TO 3C286 (S = 7.41 Jy at 4.87 GHz)

0917 + 6250	FEB/81	1.322	
1044 + 7119	APR/82	0.913	
	MAY/82	0.938	
	JUN/82	0.920	<S> = 0.934 + 0.019 Jy
	AUG/83	0.961	
	OCT/83	0.938	

TABLE 2

Flux Densities and Their Estimated Errors for the 10 Brightest Discrete Radio Sources in M82 at 5 Different Epochs

SOURCE NAME	FLUX DENSITY IN FEB '81 (Janskys)	FLUX DENSITY IN APRIL '82 (Janskys)	FLUX DENSITY IN JUNE '82 (Janskys)	FLUX DENSITY IN AUG '83 (Janskys)	FLUX DENSITY IN OCT '83 (Janskys)	% CHANGE	EXPONENTIAL HALF LIFE (Years)
39.1 + 57 3	4.30 ± 0.22	4.36 ± 0.22	4.19 ± 0.22	3.94 ± 0.21	3.50 ± 0.21	-18.6 ± 7.9	9.1
40.7 + 55 0	7.39 ± 9.25	6.86 ± 0.24	6.90 ± 0.24	6.83 ± 0.24	6.75 ± 0.24	- 8.7 ± 4.9	20.7
41.3 + 59 6	3.28 ± 0.21	3.35 ± 0.21	3.24 ± 0.21	2.62 ± 0.21	3.11 ± 0.21	- 5.2 ± 9.3	35.3
41.5 + 59 7	7.07 ± 0.24	<1.5					
41.9 + 58	108.54 ± 2.18	93.95 ± 1.89	91.65 ± 1.84	85.16 ± 1.71	79.49 ± 1.60	-26.8 ± 2.8	6.0
43.2 + 58 3	5.18 ± 0.23	4.18 ± 0.37	4.76 ± 0.22	4.35 ± 0.22	4.12 ± 0.22	-20.5 ± 6.9	8.2
43.3 + 59 1	10.92 ± 0.30	10.21 ± 0.29	9.55 ± 0.28	10.10 ± 0.28	9.56 ± 0.28	-12.4 ± 4.0	14.1
44.0 + 59 5	24.22 ± 0.52	22.60 ± 0.49	22.11 ± 0.49	22.37 ± 0.49	21.22 ± 0.47	-12.4 ± 3.1	14.2
45.2 + 61 2	7.52 ± 0.25	7.47 ± 0.25	6.70 ± 0.24	6.96 ± 0.24	6.52 ± 024	-13.3 ± 5.0	13.2
45.7 + 65 2	2.55 ± 0.21	2.40 ± 0.21	2.57 ± 0.21	2.89 ± 0.21	2.26 ± 0.21	-11.4 ± 12.4	15.6

IV. RESULTS OF LUMINOSITY MONITORING OVER 2.7 YEARS

Table 2 gives the flux densities for the brightest ten sources in February 1981 at five observing epochs, and the percentage change in flux by comparing the beginning and end epoch fluxes. This is converted to a half-life, assuming an exponential rate of decay.

For fainter sources, the flux densities become proportionately more uncertain owing to the residual lumpy, complex background of continuum emission. For sources ranging from 2.5 mJy to 1 mJy we would expect fictitious changes in flux due to measurement errors of 8% to 20% respectively. This effect can be seen in Table 3 which compares the "end-point" fluxes for the 9 next brightest sources, whose flux densities lie between 1.4 and 2.5 mJy at the beginning epoch. However, in spite of the larger percentage errors due to low-level systematic effects and noise, 7 of the 9 less luminous sources have changed by less than 10%. By comparison, only 2 of the brightest ten sources (Table 2) decreased by less than 10%. Thus, even without deconvolving the errors, it is clear that the fainter sources are, statistically, more stable than the brightest ones. This result is consistent with expectation that the fainter ones are on average older and were, just a few years ago, among the brightest ones which have decayed in a power-law, or exponential fashion.

TABLE 3

Comparison of the 1983 October 5 GHz Flux Densities with Those in 1981 February from Kronberg, Biermann and Schwab (1985), for the 9 next Brightest Radio Sources in M82, which have Flux Densities Between 1.4 and 2.5 mJy

SOURCE NAME	Flux Density in Feb. '81 (Janskys)	Flux Density in Oct. '83 (Janskys)	% CHANGE	EXPONENTIAL HALF-LIFE (years)
40.6 + 56 0	1.40 ± 0.20	1.43 ± 0.20	2.1 ± 20.00	
42.2 + 59 0	1.49 ± 0.28	1.68 ± 0.20	12.7 ± 22.2	
42.7 + 55 6	1.75 ± 0.20	1.75 ± 0.20	0.0 ± 16.2	
44.3 + 59 2	2.51 ± 0.21	2.36 ± 0.21	-6.00 ± 12.2	30.5
44.5 + 58 1	2.10 ± 0.20	2.73 ± 0.21	30.00 ± 12.2	
44.9 + 61 1	1.82 ± 0.20	1.69 ± 0.20	-7.14 ± 16.2	25.4
45.4 + 67 3	1.49 ± 0.20	1.61 ± 0.20	8.05 ± 18.3	
45.9 + 63 8	1.96 ± 0.20	1.93 ± 0.20	-1.53 ± 14.5	121.9
46.5 + 63 8	2.27 ± 0.21	2.15 ± 0.20	-5.29 ± 13.1	34.6

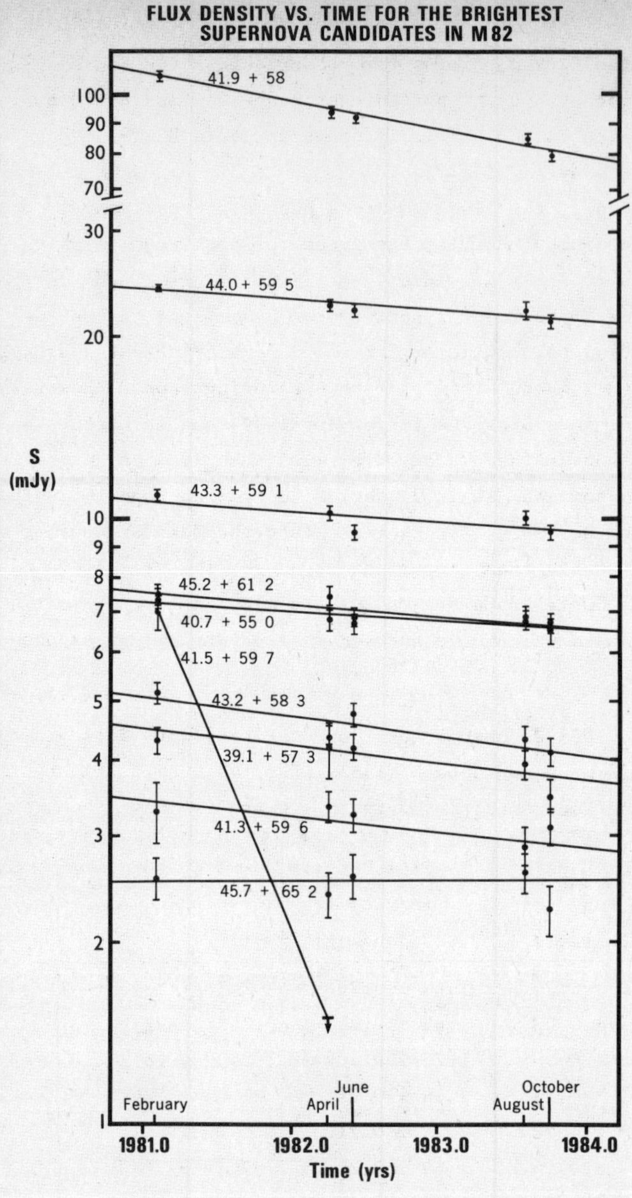

FIGURE 2. Flux density-time plots for the ten M82 sources which were stronger than 2.5 mJy in February 1981, as measured between 1981 February and 1983 October. For each source (except for 41.5 + 597) a least squares straight line best fit is shown.

Longer monitoring times are required to establish the character of the radio luminosity decay curves, i.e. whether they are power law, exponential, etc. and to better establish differences in the decay characteristics among the many radio supernovae. In Figure 2 (reproduced from Kronberg and Sramek, 1984), we show flux density-time plots for the ten sources which were stronger than 2.5 mJy in 1981 February. Sources are labelled according to their positions, as indicated in the contour map at epoch 1981.1 (Figure 3) (cf. Kronberg, Biermann and Schwab, 1985). With the exception of 41.3 + 59 6, the brightest 10 have all decreased, with half-lives which are less than ~20 years.

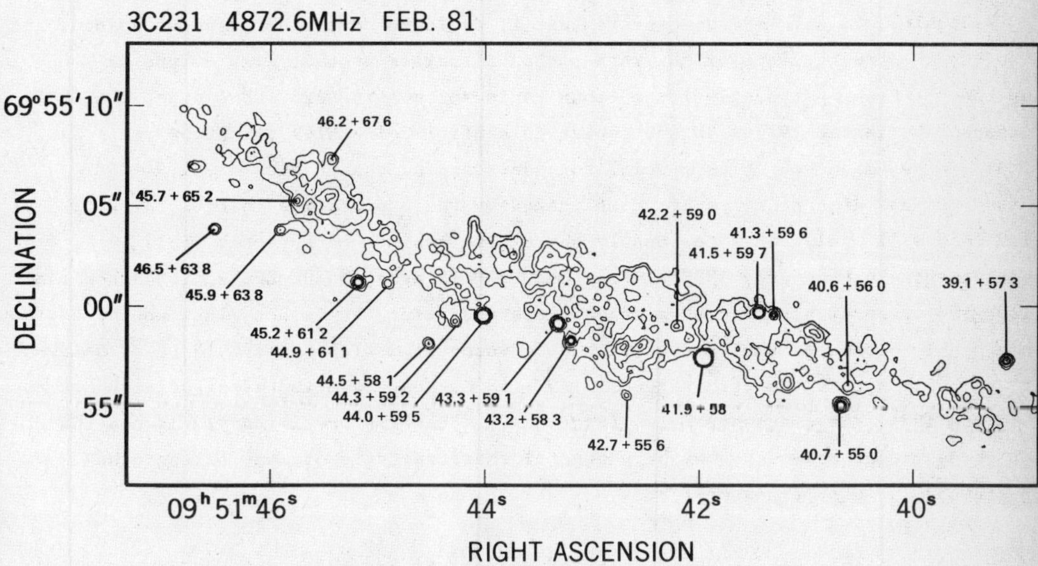

FIGURE 3. Radio map of the inner 600 pc of M82, made with the NRAO VLA at 0.34" resolution. The brightest point source is 41.9 + 58. The faintest sources visible have radio luminosities comparable to that of Cassiopeia A, if it were at the distance of M82 (3.2 Mpc). Contours are shown at 5, 10, 20, 30, and 40% of peak surface brightness. The sources labelled are those listed in Tables 2 and 3.

We describe the flux density variations of two noteworthy sources in more detail:

(i) The Brightest Source, 41.9 + 58.

This source has decreased at an average rate of 9%/yr from 1981.1 to 1983.8. This agrees well with the average rate when earlier 6 cm data at each epoch 1973.6 (Hargrave, 1974) are included, and also 1971 measurements at 8 GHz (Kronberg and Clarke, 1978). The agreement between the rate of decrease of 41.9 + 58 in our present data and that in previous data above $\sim$2 GHz gives further confidence to our flux density calibration, and hence the variability estimates for other sources having flux densities which are well above the noise and radio background levels.

The nature of 41.9 + 58 has been discussed in several papers (Geldzahler et. al., 1977, Kronberg and Clarke, 1978, Brown and Neff, 1980, Kronberg, Biermann and Schwab, 1981, 1985), and its nature remains uncertain. Its flux decay characteristics and luminosity are comparable with other sources in Table 2, and radio supernovae in other galaxies. On the other hand, its VLBI-measured size of only 1.5 milliarcseconds by Geldzahler et. al. (1977) suggests that it could be some other type of object, and possibly the true nucleus of M82 rather than a stellar system, e.g. a Type II supernova, which is typical of the other radio sources in M82. Some other VLBI measurements in the 1970's do not appear to confirm Geldzahler et. al.'s small size -- a parameter which is crucial to understanding the physics of 41.9 + 58. These apparent disagreements are complicated by the expectation that the size of 41.9 + 58 will evolve on a reasonably short time scale. More recent VLBI measurements in the early 1980's described by Wilkinson and de Bruyn (1984) now show 41.9 + 58 to be highly aligned along a preferred axis. It is not clear whether or not this highly elongated VLBI structure can be reconciled with the 1974 data of Geldzahler et. al., given that the structure of 41.9 + 58 may have evolved substantially between 1974 and 1981. As discussed below (Section V), there is preliminary evidence that a high degree of asymmetry may be a general characteristic of many of the other (presumed) radio supernovae in M82.

(ii) Discovery of a "Rapid Turnoff" Source, 41.5 + 597.

The most dramatic change has occured for source 41.5 + 597, which, in February 1981 was the sixth brightest source at 7.1 mJy. Fourteen months later, and in all subsequent sessions it was less than 1.5 mJy. It had a spectral index of -1.1 in February 1981, and was unresolved at both 4.9 and 15 GHz, which gives a firm size upper limit of 0.15". It is clear that the flux density variation of this source is distinct from all the others, which suggests that it arises in a different kind of object.

It is interesting to compare 41.5+597 in M82 with the supernova 1983.51 observed in M83 by Sramek, Panagia and Weiler (1983). This object had a peak radio luminosity of 10^{20} WHz^{-1}, which is comparable to our value of 1.1x10^{19} WHz^{-1} measured for 41.5+597 in 1981 February. Also, Sramek et al. found SN 1983.51 to have a decay time of less than 100 days, which makes its behavior consistent with 41.5+597 in M82. The spectral index between 5 and 14.6 GHz of 41.5+597 in February 1981 was -1.1 (Kronberg, Biermann and Schwab, 1985). Thus, the luminosity, decay time, and spectral index of 41.5+597 are all similar to SN 1983.51 in M83. The rapid decay time of 41.5+597 suggests in any case that its first observation in 1981.1 was very close in time to the supernova event.

Our monitoring program is continuing. Attempts are also continuing to improve the dynamic range and image quality of the 6cm maps. As these improve, we hope to obtain a more consistent definition of the background radio emission, and hence improved precision in determining flux densities free of systematic low-level background effects. Also, with further improved dynamic range, it should be possible to detect more than the current ∿40 discrete sources in our 6 cm maps.

V. ANGULAR SIZE AND STRUCTURE OF THE M82 RADIO SUPERNOVAE

The brightest sources are either unresolved or slightly resolved by the 0.3" beamwidth of the VLA at 6cm. This corresponds to a linear size of 5 pc at an assumed distance of 3.2 Mpc for M82. The partially resolved appearance of some sources is confirmed by our VLA observations at 2cm, which show some, such as 41.9+58 and 41.5+597 to be unresolved (<1.5 pc), but several others to have elongated structure on size scales of 1-4 pc. None of the resolved sources shows a symmetrical structure. On the contrary, they all show a marked asymmetry. This is a most interesting result, which will be better defined when we obtain our combined configuration VLA maps at 2cm and 1.3cm wavelength.

Our preliminary values for the angular sizes (see also Kronberg, Biermann and Schwab, 1985), combined with previous VLBI measurements for 41.9+58 (Geldzahler et al., 1977, Wilkinson and de Bruyn, 1984), show that the sources in Tables 2 and 3 range in angular size from a few to ∿500 milliarcseconds. The implications of the asymmetrical structures for using radio SN as distance indicators are not yet entirely clear. However if highly asymmetrical structures are common, we will not be able to straightforwardly apply methods which assume spherically expanding radiating shells. Radio emitting jets may be involved, a possibility which was suggested by Kronberg, Biermann and Schwab (1985).

VI. IMPLICATIONS FOR USING SUPERNOVAE AS DISTANCE INDICATORS

These preliminary results on size, when combined with our new variability results also give us some first estimates for the expansion rates, in particular the sources which extend up to 4-5 pc. If we adopt an age upper limit of 100 years for a source of 300 milliarcseconds extent, it suggests that at least some radio emitting clouds must be expanding in their longest dimension by $\gtrsim$22,000 km/s. Generalized to other galaxies, this implies an angular rate of size growth is $\geq 0.96/D_{10}$ milliarcseconds/year where D_{10} is the galaxy's distance in units of 10 Mpc. -- i.e. $\sim$1 mas/year at the distance of the Virgo Cluster.

Preliminary as they are, our angular size results are intriguing. They suggest that the size growth of radio supernovae is (1) highly asymmetrical, and (2) rather rapid. If characteristic (1) proves to be typical of radio supernovae, this may limit their usefulness as distance indicators. They may also be indicating that some radio supernovae form bipolar jets. There is the additional possibility that at least some of the objects in M82 may have SS433-like properties which produce unusually high radio luminosities in the dense interstellar environment of M82's nucleus. If the inferred rapid expansion is a common characteristic, it may permit us to determine statistical expansion velocities scaled to the distance of M82. This, if transferrable to other active galaxies in the nearby universe, might provide a distance scale indicator.

VII. CONCLUSIONS

The phenomenon of rapid variability affords us the first opportunity of directly estimating the rate of energy input to the interstellar medium of an active galaxy. We have discovered that virtually all of the brightest radio sources in M82 are decreasing in luminosity on the remarkably short time scale of a few years. At the current rates of decrease, eight out of the 10 brightest sources in Figure 1 which were brighter than 3 millijansky in early 1981 will be fainter than this level in less than $\sim$35 years!

At least one source, 41.5+597, has a remarkably fast decay time of less than one year, whereas another, comparably bright one, 45.2+612, has been uncommonly stable over the 2.7 year period. These facts reveal that individual variability behavior, and perhaps even the peak radio luminosities within our population can be widely different, and any modelling of this dynamically evolving population must account for this fact. The relative occurence rates of fast (Type I SN?), and slow decayers has yet to be established. Further monitoring should clarify this.

To maintain such a rapidly decreasing population of luminous sources, our results require a refreshment rate of one new radio source, i.e., presumably a new supernova, every $\sim$4.5 years. The very rapid decay of source 41.5+597 strongly suggests that its first observation in February 1981 was very close to the supernova outburst.

ACKNOWLEDGEMENTS

This research was supported in part by the Natural Sciences and Engineering Research Council (NSERC) of Canada. I thank my collaborators, P. Biermann, F.R. Schwab, and R.A. Sramek, who co-authored the two recent papers describing most of these results, and also L. Fenton-Lloyd and B. Glendenning for their assistance with the data reduction.

REFERENCES

Brown, R.L., and Neff, S.G., 1980, Ap. J. 241, 561

Hargrave, P.J., 1974, M.N.R.A.S., 168, 491

Geldzahler, B.J., Kellermann, K.I., Shaffer, D.B., and Clark, B.G. 1977, Ap. J. (Letters), 215, L5

Kronberg, P.P., and Biermann, P. 1983, Int'l Astron. Union Symposium No. 101., 583

Kronberg, P.P., and Clarke, J.N., 1978, Astrophys. J. (Letters), 224, L51.

Kronberg, P.P., Biermann, P. and Schwab, F.R., 1981, Astrophys. J., 246, 751. (KBS1)

ibid., 1985, Astrophys. J., (in press). (KBS2)

Kronberg, P.P. and Sramek, R.A., 1984, Science (in press)

Sramek, R.A., Panagia, N., and Weiler, K.W., 1984 (in press)

Weiler, K.W., Sramek, R.A., van der Hulst, J.M., and Panagia, N., 1983, Radio Supernovae, IAU Sympoisum 101, page 171.

Wilkinson, P.N. and de Bruyn, G., 1984, M.N.R.A.S. (in press).

Detecting Supernova Remnants in External Galaxies

John R. Dickel
Astronomy Department, University of Illinois

and

Sandro D'Odorico
European Southern Observatory

Abstract

Reliable identifications of supernova remnants in external galaxies are slowly becoming available through the combined use of radio, optical, and x-ray surveys. The host galaxies represent a variety of Hubble types. Although the number of SNR detected in each galaxy is currently small, new telescopes and detectors in all three wavelength regimes should allow us to obtain larger samples in the near future.

I. Introduction

The radiation from supernovae (SN) is generally thought to arise in the outer layers of the exploding stars whereas the radiation from what I shall call supernova remnants (SNR) is caused by the interaction of the expanding material and shocks with their surroundings. The transition between the SN and SNR phases will be controlled largely by the circumstellar medium but also by the mass distribution and energy of the SN itself. To most effectively investigate these properties we need to observe an SN become an SNR. However, historical SN in external galaxies have been recorded for only about 100 years and none of these has yet produced an observable SNR (de Bruyn 1973; Brown and Marscher 1978; Ulmer et al. 1980; Cowan and Branch 1982; Dickel and D'Odorico 1984). This is in general accord with the theories which predict that SNR should turn on approximately 100 years after explosion (e.g. Gull 1973) although Cowsik and Sarkar (1984) indicate that some may begin in a time as short as 30 years. Thus it is important to periodically check the positions of historical SN in other galaxies for evidence of the beginning of SNR emission but we also need catalogs of SNR identified by other means. Such information should be obtained in a variety of galaxies so that with sufficient statistics it may be possible to relate the initial stellar populations and their supernova production to the galactic environments controlling the expansion of their SNR.

II. Present Results

Reasonably complete catalogs currently exist for the Milky Way (Green 1984) and the Magellanic Clouds (Mathewson et al. 1984) but results to date for other galaxies are very incomplete. A few remnants have been detected in M31 (D'Odorico, Dopita, and Benvenuti 1980, hereafter DDB; Dickel and D'Odorico 1984) and M33 (DDB; D'Odorico et al. 1982), one or more in M82 (Kronberg and Biermann 1983; Kronberg, this workshop), and single objects in the irregular galaxy NGC 4449 (Balick and Heckman 1978; Seaquist and Bignell 1978), the dwarf irregular NGC 6822 and IC 1613 (D'Odorico and Dopita 1983; Dickel, D'Odorico, and Silverman 1985), and probably the dwarf elliptical NGC 185 (Gallagher, Hunter, and Mould 1984). Two objects in the spiral NGC 5236 might possibly be SNR (Cowan and Branch 1982). The small numbers appear to be due largely to the sensitivity and resolution limits of present equipment but could also reflect some intrinsic differences among the objects in different galaxies and/or selection criteria.

In the Milky Way most of the 140-odd remnants known were first cataloged as radio objects. They are identified by their non-thermal spectrum with an index averaging about -0.5 and shell structures (or flatter non-thermal spectrum and smooth filled structure for the Crab-type remnants). Less than 1/5 of them have any optical counterparts (van den Bergh, Marscher, and Terzian 1973; van den Bergh 1978) and in several cases only a small part of the shell shows optical features. A good example of this was recognized by Hill (1967) who identified the optical feature RCW 86 (Rodgers, Campbell, and Whiteoak 1960) with the southwestern clump of the radio SNR MSH 14-6$\underline{3}$ (Mills, Slee, and Hill 1961). An overlay of a more recent radio map of the area (Milne and Dickel 1975) on the photograph from the SRC survey is shown in Figure 1. Many of the SNR in the Milky Way have not yet been fully investigated at x-ray wavelengths so meaningful statistical comparisons employing that wavelength regime are not yet available.

In the Magellanic Clouds the most complete lists of SNR come from the x-ray surveys with the Einstein Observatory (Long, Helfand, and Grabelsky 1981; Seward and Mitchell 1981; Inoue, Koyama, and Tanaka 1983). Most of these have now been mapped at radio wavelengths using the Molonglo Observatory Synthesis Telescope (Mills et al. 1984). Again, many do not show optical emission.

In the next section we shall describe the procedures for obtaining lists of the SNR in other galaxies.

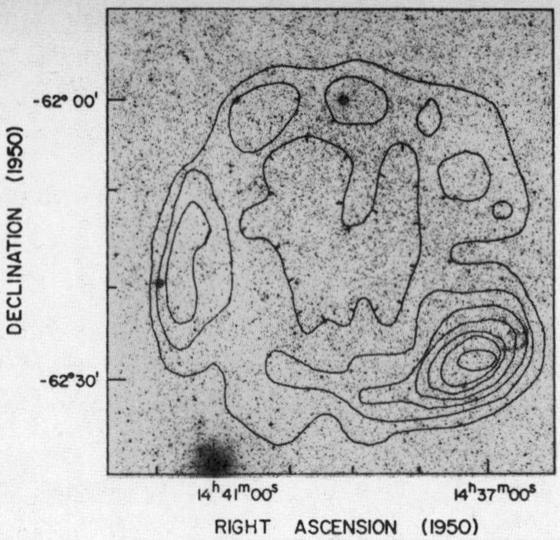

Figure 1. A radio contour map at 6 cm of the supernova remnant MSH 14-6[3] superimposed upon the optical photo from the SRC Southern Hemisphere Survey. The half-power beamwidth of the radio observations was 4.4 arcmin and the contour interval is 0.1 K in brightness temperature. The only prominent optical emission arises in the small arc somewhat above and to the right of the brightest radio emission in the southwest; there is not a close correlation of features. A very faint optical filament is also present across part of the northern part of the shell.

III. Techniques

A. Radio Surveys

It would appear on the above basis that the best way to identify and catalog SNR is by radio surveys. Some of the missing objects at optical and x-ray wavelengths can be accounted for by obscuration. However, as described by Blandford and

Cowie (1982) there can also be radio emission with little or no optical or x-ray. The radiation from an SNR arises in density fluctuations in the surrounding medium which have been heated and compressed by the expanding shock and ejectum. Once the radiative shocks have crossed a cloud the x-ray emission from the heated gas and the optical recombination lines from its subsequent cooling will quickly decay, leaving only the synchrotron emission from the relativistic particles with long radiation lifetimes.

There are severe problems with the radio surveys of other galaxies, however. Until recently sensitivity and resolution were important restrictions but these have been largely relieved by the VLA. Now confusion and contaminations by background objects in the field cause the most serious problem. Adopting a reasonable cutoff in flux density of 1/2 mJy for VLA surveys at 20 cm, we can estimate the expected number of SNR above this limit based upon the statistics of galactic remnants. In M31, for example, if the remnants have the same luminosity as those in the Milky Way and a similar number-diameter relation then we should expect fewer than 100 SNR brighter than 1/2 mJy at 20 cm in that galaxy. Extrapolations of statistical counts of background sources (Oosterbaan 1978; Fomalont, Bridle, and Davis 1974) predict about 1000 background sources brighter than this limit within the boundary of M31. Thus only about one-tenth of the sources found would be the desired candidates and there are no good ways to distinguish which is which. Although the SNR in galaxies of the Local Group will have diameters of typically several arcseconds, so do many unidentified sources. Both kinds of objects have significantly polarized emission and a large range in non-thermal spectral indices. An example of confusing sources around an SNR in M31 is illustrated by the VLA map shown in Figure 2 (from observations by Dickel and D'Odorico 1984). The SNR is the small arc-like structure covering about 10 arcsec right in the center of the field but several other equal or brighter sources are also present. They have not been identified. Much detailed follow-up work using other techniques will be necessary to distinguish extragalactic SNR from the list of candidates provided by radio surveys.

B. Optical Surveys

The optical line emission from SNR shows very characteristic intensity ratios among various lines caused by the shock excitation of the emitting gas. Although many ratios are dependent upon abundances and physical conditions (e.g. Raymond 1979) the ratio of the [S II] doublet at .6717 and .6731 μm to Hα is an excellent discriminant of SNR -- it is greater than 0.4 for SNR and less than that value for H II regions (see e.g., D'Odorico 1978). This criterion was used to successfully identify objects in the LMC (Mathewson and Clarke 1973) and other galaxies (DDB).

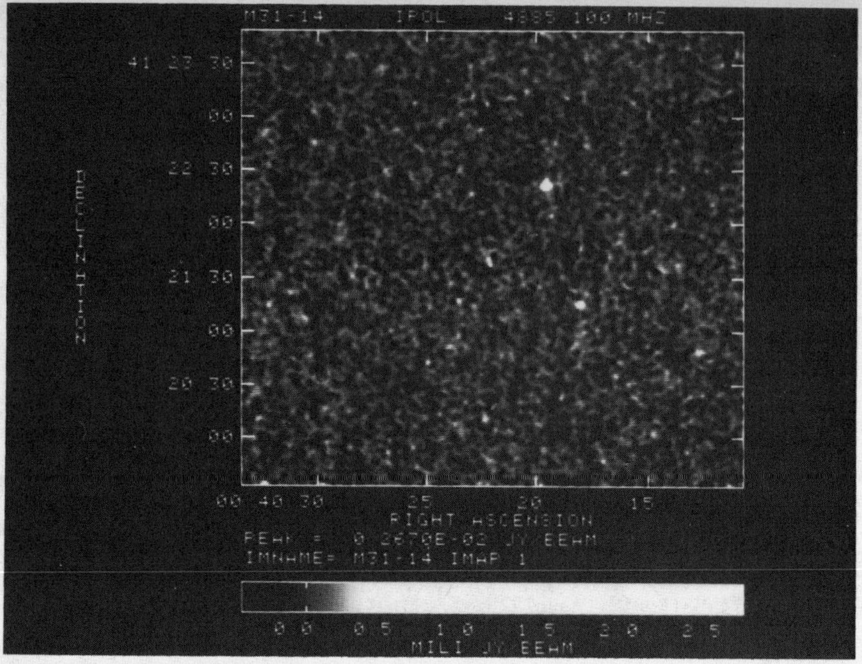

Figure 2. 6-cm radiophoto of a small field in M31 showing one SNR and several unidentified contaminating sources. The half-power beamwidth was 3 arcsec.

The general procedure is to compare filter photographs in the two lines and then do follow-up spectroscopy or radio studies to confirm the identification. In a number of instances, however, no radio source has been found at the position of an optical feature with a confirmed high [S II]/Hα ratio. Upper limits to the radio surface brightness are often less than one-tenth the value expected for remnants of similar size in the Milky Way (e.g. Dickel and D'Odorico 1984; Dickel, D'Odorico, and Silverman 1985). Most are probably real SNR although a few might be H II regions with peculiar shock excitation from stellar winds (Lasker 1977). Either the optical features cover only a small part of a given remnant so that it is larger and thus the resultant radio surface brightness is lower than predicted (as would happen for MSH 14-6$\underline{3}$ in Figure 1) or these optically detected objects may fall largely into a class of radio quiet SNR mentioned by Blandford and Cowie (1982).

A major limitation of this criterion employing the [S II]/Hα ratio is that it has not found many SNR. DDB listed only 17 confirmed objects in M31. In a new study of that galaxy, R. Walterbos (private communication) is digitizing and machine comparing plates in Hα, [S II], and a blue continuum taken by G. de Bruyn. He is finding a number of new candidates but a spectroscopic check of a few of these by us has suggested that many may be stellar images or plate irregularities. Complete confirmatory observations are needed but we suspect that the final list will be shorter than that of the SNR in our own Galaxy.

C. X-ray Surveys

Most of the SNR in the Magellanic Clouds have been detected as x-ray emitters (Long, Helfand, and Grabelsky 1981; Seward and Mitchell 1981; Inoue, Koyama, and Tanaka 1983) but toward M31, which is of order 10 times more distant, there are no coincidences between the lists of optical SNR (DDB) and x-ray sources (van Speybroeck et al. 1979). We must await the increased sensitivity of AXAF to detect these objects in distant galaxies. In addition, many x-ray sources are point-like binary stars and it will be necessary to have high angular resolution to separate the slightly extended SNR from the binaries.

IV. Discussion

It appears that a good identification of an extragalactic SNR generally requires an iterative procedure of observations in more than one wavelength regime. The process is often time consuming and many candidates are not confirmed. Lest this report seem pessimistic, however, let us point out that progress is certainly being made. Only six years ago, no confirmed SNR were known in any galaxy beyond the Milky Way and the Magellanic Clouds. Since then we have approximately doubled the number in the Magellanic Clouds and found about two dozen likely SNR in more distant galaxies. During the next decade, with the continued use of the VLA plus the Space Telescope and AXAF, we should see a dramatic increase in these numbers. This will give a meaningful sample and allow us to compare extragalactic supernova remnants and their statistics with those of extragalactic supernovae and also with the Galactic samples.

Acknowledgements

We thank Eric Jones for useful comments. Some of this research has been supported in part by NASA.

References

Balick, B. and Heckman, T. 1978. Ap. J. Lett., **226**, L7.

Blandford, R. D. and Cowie, L. L. 1982. Ap. J., **260**, 625.

Brown, R. L. and Marscher, A. P. 1978. Ap. J., **220**, 467.

Cowan, J. J. and Branch, D. 1982. Ap. J., 258, 31.

Cowsik, R. and Sarkar, S. 1984. Monthly Not. Royal Astron. Soc., 207, 745.

de Bruyn, G. 1973. Astron. Astrophys., 26, 105.

Dickel, J. R. and D'Odorico, S. 1984. Monthly Not. Royal Astron. Soc., 202, 351.

Dickel, J. R., D'Odorico, S., and Silverman, A. 1985. submitted to Astron. J.

D'Odorico, S. 1978. Mem. Soc. Astron. Ital., 49, 485.

D'Odorico, S. and Dopita, M. 1983. in Supernova Remnants and Their X-ray Emission, ed. by I. J. Danziger and P. Gorenstein (Dordrecht: Reidel), p. 551.

D'Odorico, S., Dopita, M., and Benvenuti, P. 1980. Astr. Astrophys. Suppl., 40, 67.

D'Odorico, S., Goss, W. M., and Dopita, M. 1982. Monthly Not. Royal Astron. Soc., 198, 1059.

Fomalont, E. B., Bridle, A. H., and Davis, M. M. 1974. Astron. Astrophys., 36, 277.

Gallagher, J. S., Hunter, D. A., and Mould, J. 1984. Ap. J. Lett., 281, L63.

Green, D. 1984. Monthly Not. Royal Astron. Soc., 209, 449.

Gull, S. 1973. Monthly Not. Royal Astron. Soc., 161, 47.

Hill, E. R. 1967. Australian J. Phys., 20, 297.

Inoue, H., Koyama, K., and Tanaka, Y. 1983. in Supernova Remnants and Their X-ray Emission, ed. by I. J. Danziger and P. Gorenstein (Dordrecht: Reidel), p. 535.

Kronberg, P. 1984. this workshop.

Kronberg, P. and Biermann, P. 1983. in Supernova Remnants and Their X-ray Emission, ed. by I. J. Danziger and P. Gorenstein (Dordrecht: Reidel), p. 583.

Lasker, B. M. 1977. Ap. J., 212, 390.

Long, K. S., Helfand, D. J., and Grabelsky, D. A. 1981. Ap. J., 248, 925.

Mathewson, D. S. and Clarke, J. N. 1973. Ap. J., 180, 725.

Mathewson, D. S., Ford, V. L., Dopita, M. A., Tuohy, I. R., Mills, B. Y., and Turtle, A. J. 1984. Ap. J. Suppl., 55, 189.

Mills, B. Y., Slee, O. B., and Hill, E. R. 1961. Australian J. Phys., 14, 497.

Mills, B. Y., Turtle, A. J., Little, A. G., and Durdin, J. M. 1984. Australian J. Phys., 37, 321.

Milne, D. K. and Dickel, J. R. 1975. Australian J. Phys., 28, 209.

Oosterbaan, K. 1978. Astron. Astrophys., 69, 235.

Raymond, J. C. 1979. Ap. J. Suppl., 39, 1.

Rodgers, A. W., Campbell, C. T., and Whiteoak, J. B. 1960. Monthly Not. Royal Astron. Soc., 121, 103.

Seaquist, E. R. and Bignell, R. C. 1978. Ap. J. Lett., 226, L5.

Seward, F. D. and Mitchell, M. 1981. Ap. J., 243, 736.

Ulmer, M. P., Crane, P. C., Brown, R. L., and van der Hulst, J. M. 1980. Nature, 285, 151.

van den Bergh, S. 1978. Ap. J. Suppl., 38, 119.

van den Bergh, S., Marscher, A. P., and Terzian, Y. 1973. Ap. J. Suppl., 26, 19.

van Speybroeck, L., Epstein, A., Forman, W., Giacconi, R., Jones, C., Liller, W., and Smarr, L. 1979. Ap. J. Lett., 234, L45.

ANGULAR DIAMETER DETERMINATIONS OF RADIO SUPERNOVAE AND THE DISTANCE SCALE

Norbert Bartel

Harvard-Smithsonian Center for Astrophysics
Cambridge, MA 02138, USA

ABSTRACT

A new method of determining extragalactic distances and inferring H_o is presented. The method combines the determinations of radial expansion velocities of supernovae via optical spectroscopy and the determinations of angular expansion velocities via Very Long Baseline Interferometry (VLBI). The data of two recent supernovae, SN 1979c in the galaxy M100 and SN 1980k in the galaxy NGC 6946, have been analyzed. We obtained an estimate of the distance, D, to M100 of $11.4\alpha_\nu \lesssim D$ [Mpc] $\lesssim 24.4\alpha_\nu$ and a lower limit of the distance to NGC 6946 of $D > 2.3\alpha_\nu$ [Mpc] with the factor α_ν coupling the expansion velocities of the "radiosphere" and the photosphere and being most likely $\gtrsim 1$. These distances are equivalent to $100\alpha_\nu^{-1} \gtrsim H_o$ [km s^{-1} Mpc^{-1}] $\gtrsim 40\alpha_\nu^{-1}$ and $H_o < 230\alpha_\nu^{-1}$ km s^{-1} Mpc^{-1} for the two measurements, respectively.

I. INTRODUCTION

In Figure 1 I display a map of the galaxy M100 obtained with the Very Large Array[1] (VLA) in the D array at 5.0 GHz on 1982 Dec. 8. The supernova SN 1979c is situated along the southeast direction $\sim$ 100 arcsec away from the nucleus at the southern edge of a spiral arm. The flux density of the supernova at 5 GHz reached a maximum of $\sim$ 8 mJy in the early months of its evolution and subsequently decreased to a present (end of 1984) level of $\sim$ 4 mJy (see Weiler 1985). Optical spectroscopic observations indicate that SN 1979c is expanding with a maximum velocity of $\sim$ 10500 km s^{-1} (Branch *et al.* 1981). If we assume uniform expansion and if the distance of M100 is between 11.5 Mpc (de Vaucouleurs 1975) and 22.2 Mpc (Sandage and Tammann 1979), the increase of the angular diameter of the supernova is 0.39 to 0.20 mas per year, respectively. Such a rate of change can be monitored when the most sensitive interferometers with the longest baselines presently available are used. The same is true for most supernovae that exhibit at radio frequencies $\sim$ 1 GHz flux densities of $\gtrsim 2$ mJy over at least a few years.

---

[1] The VLA is operated by the National Radio Astronomy Observatory through the Associated Universities, Inc., under contract to the National Science Foundation.

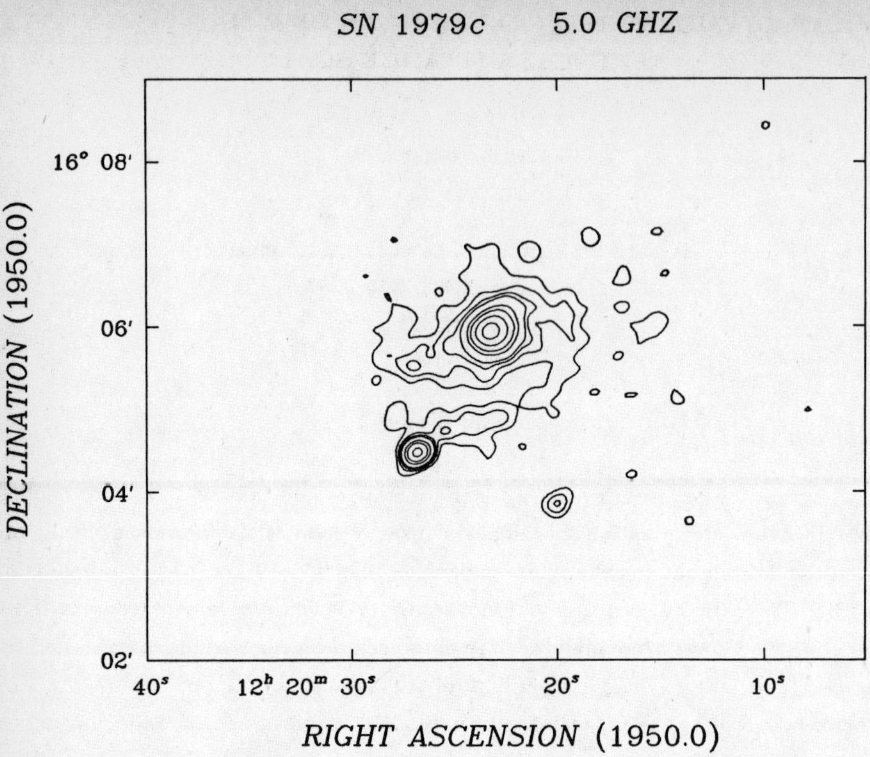

Figure 1. Radio map of the supernova SN 1979c in the galaxy M100. The contour levels are at 2.5, 5.0, 7.5, 10, 20, 30, 50, and 80 percent of the peak flux density per beam area.

When such VLBI determinations of angular expansion velocities of supernovae are coupled with spectroscopic determinations of radial expansion velocities, the distances of supernovae can be derived and H_o inferred. Here I present and discuss the first such determinations.

II. VLBI OBSERVATIONS OF THE SUPERNOVAE SN 1979c AND SN 1980k

We made VLBI observations of SN 1979c at several epochs after the time of explosion, which is assumed to be 1979 Apr. 1, about 19 days before maximum light. We also observed SN 1980k at two epochs after its time of explosion, which I assume to be 1980 Oct. 17, also about 19 days before maximum light. The observation series is summarized in Table 1.

At each station a hydrogen maser frequency standard was used to govern the local-oscillator chain and the "time tagging" of the recorded data. All observations were made with the Mark III VLBI system in recording mode A using a bandwidth of ~ 56 MHz (see Rogers *et al.* 1983). The data were correlated with the Mark III VLBI processor of the Haystack Observatory located in

Table 1
Observations

Source[1]	Date of Observation	Frequency (GHz)	Polarization[2]	Antennas[3]	MSIA[4]	Sensitivity[5] of MSIA (7σ) (mJy)	No. of scans[6] with MSIA
SN 1979c	1982 Dec. 8	5.0	LCP	BGOY	BY	1.7	4
SN 1979c	1983 May 9	8.3	RCP	(BDKM)	DM	2.1	2
SN 1979c	1983 May 9	2.3	RCP	BDFG(K)MO(S)	DM	1.5	2
SN 1979c	1983 Dec. 1	5.0	LCP	BG(K)OY	BY	1.7	7
SN 1979c	1984 Apr. 2	5.0	LCP	G(O)Y	GY	3.6	4
SN 1979c	1984 May 31	1.7	LCP	BGOY	BY	1.4	5
SN 1980k	1983 May 7	2.3	RCP	(B)DM	DM	1.5	3
SN 1980k	1984 May 31	1.7	LCP	(BOY)	BY	1.4	1

[1] According to Weiler *et al.* (1983), SN 1979c: $\alpha(1950.0) = 12^h20^m26\overset{s}{.}71$, $\delta(1950.0) = 16°04'29\overset{''}{.}5$; SN 1980k: $\alpha(1950.0) = 20^h34^m26\overset{s}{.}68, \delta(1950.0) = 59°55'26\overset{''}{.}5$.

[2] LCP: Left-hand circular polarization.
RCP: Right-hand circular polarization.

[3] B: 100 m-antenna in Effelsberg near Bonn, F.R.G., belongs to the Max-Planck-Institut für Radioastronomie.
D: 64 m-antenna in Goldstone, CA, U.S.A., belongs to NASA.
F: 25 m-antenna in Ft. Davis, TX, U.S.A., belongs to Harvard College Observatory.
G: 43 m-antenna in Green Bank, WV, U.S.A., belongs to the National Radio Astronomy Observatory.
K: 36 m-antenna in Westford, MA, U.S.A., belongs to the Northeast Radio Observatory Corporation.
M: 64 m-antenna near Madrid, Spain, belongs to NASA.
O: 40 m-antenna near Big Pine, CA, U.S.A., belongs to California Institute of Technology.
S: 25 m-antenna in Onsala, Sweden, belongs to Onsala Space Observatory.
Y: Equivalent to a ~ 130 m-antenna ($\sim 26 \times 25$ m antennas of the VLA "phased up") near Socorro, NM, U.S.A., belongs to the National Radio Astronomy Observatory.
All 2-element interferometers with the symbol for at least one antenna in parentheses failed to yield fringes. The search for fringes at 8.3 GHz is not yet completed.

[4] Most sensitive 2-element interferometer of array.

[5] Maximum sensitivity for coherent integration of 12 min and a bandwidth of 50 MHz. Here and hereafter, 1σ denotes a statistical standard deviation.

[6] Using an observing time of ~ 12 min and a bandwidth of ~ 50 MHz.

Westford, MA, U.S.A.

As supernovae are weak sources, the search for fringes was made in a way that departed from the usual procedure for strong sources. We chose fringe search windows for the single band delay, multiband delay, and fringe rate that corresponded to position windows on the sky with widths as narrow as possible but of at least an arcsecond. The window centers corresponded to the supernova positions determined previously with the VLA[1] and to the relative clock epochs determined from observations of the calibrator sources 0851+202 (OJ287) and/or 1404+286 (OQ208) that straddled the supernova observations.

When fringes were detected for at least one scan, we used the information from the observables to further narrow the fringe search windows in parameter space. In cases where fringes were detected with two 2-element interferometers of a triplet of antennas, closure observables were calculated for the third 2-element interferometer of the triplet to narrow the fringe search window to contain only one point, the closure observable. In such cases, this procedure allowed us to lower the detection threshold down to 3σ corresponding to a probability of confusion with noise of ~ 1 percent. This procedure is equivalent to the use of the global-fringe-search technique (Schwab and Cotton 1983). For observations of SN 1979c on 1983 Dec. 1, this procedure allowed us to quadruple the number of visibility points for which detection could be validly claimed. From the eight observing sessions, we had five sessions during which the supernovae were detected with 2-element interferometers with transatlantic baselines. Data from these sessions were used for further analysis.

The visibility amplitudes were segmented in time and frequency to correct for coherence losses and were then calibrated to yield correlated flux densities as described by, e.g., Bartel *et al.* (1982). The calibration coefficients were checked for our observations of the calibrator sources, which are known to have simple brightness distributions, and were found to be consistent within 10 percent, except in the case of the observations on 1984 May 31 at 1.7 GHz, for which the check has not yet been completed.

III. DETERMINATIONS OF THE ANGULAR DIAMETER OF SN 1979c

In Figure 2 we show the maximum obtainable u-v coverage for observations at 5.0 GHz with the four antennas we mainly used in our observing sessions. As SN 1979c has low declination and our VLBI array is mainly oriented eastwest, we were observing the supernova with highest resolution along the eastwest direction and with about four times smaller resolution along the

---

[1] Positions of celestial sources determined with the VLA agree with those determined with our VLBI array to within ~ 0.3 arcsec (see, e.g., Bartel *et al.* 1985 for comparison of position determinations).

northsouth direction. A similarly elongated and oriented resolution pattern was also obtained for the observations at 2.3 and 1.7 GHz.

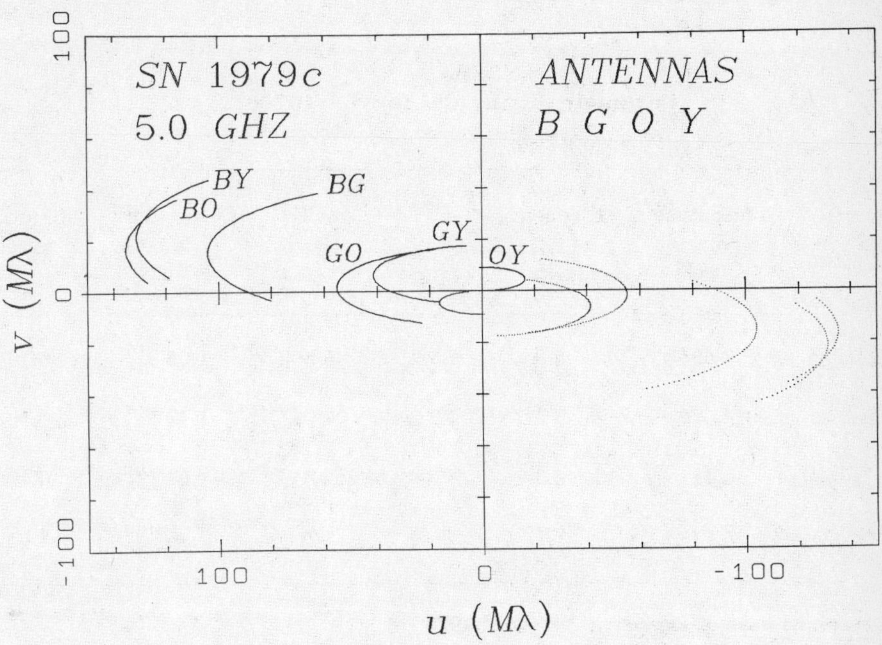

Figure 2. The maximum u-v coverage for observations of SN 1979c at 5.0 GHz with four antennas. The abbreviations for the antennas are given in Table 1. The dotted curves show reflected u-v tracks.

We estimated the parameters of a circular uniform sphere model, i.e., the total flux density and the angular diameter, by means of a weighted least-squares fit of the Fourier transform of the model to the correlated flux densities of the supernova. For the uncertainties of the correlated flux densities, we used statistical standard deviations only.

We estimated the contribution of calibration errors to the uncertainties of the parameters by changing, in turn, the sensitivity of each antenna of the array by ±15 percent while keeping the sensitivities of the other antennas fixed. For the 1.7 GHz observations, the sensitivities were varied by ±20 percent as calibration problems have yet prevented us from obtaining values for the sensitivities of higher accuracy. The largest variations in the estimated parameters due to the changes of the sensitivities of the antennas were up to two times larger than the statistical standard deviations for these parameter estimates, except for the 1.7 GHz observations, for which they were more than threefold larger. For the uncertainties of the parameters, we adopted the root-sum-squares (rss) of the values of the largest variations and the statistical standard deviations. The results of the analysis are given in Table 2.

Table 2
Parameter Estimates for SN 1979c

Date of Observation	Time since[1] Explosion (yr)	Frequency (GHz)	S_{tot}^{VLA}[2] (mJy)	S_{tot} (mJy)	Angular[3] Diameter (mas)	Reduced χ^2
1982 Dec. 8	3.691	5.0	$5.5 \pm .3$	$5.30^{+.26}_{-.54}$	$1.05^{+.09}_{-.10}$	0.70
1983 Dec. 9	4.672	5.0	$4.2 \pm .2$	$4.08^{+.45}_{-.38}$	$1.43^{+.07}_{-.07}$	0.76
1983 May 9	4.117	2.3	–	$9.7^{+.8}_{-.4}$	$1.67^{+.26}_{-.12}$	0.49
1984 May 31	5.173	1.7	8 ± 1	$8.0^{+.7}_{-.8}$	$2.36^{+.56}_{-.66}$	2.11

[1] The assumed date of explosion is 1 April 1979.
[2] The total flux density of the supernova measured with the VLA alone. For the 5 GHz data, the errors were taken to be 5 percent of the total flux density to include both statistical standard deviations and systematic errors. For the 1.7 GHz data, the error represents the variation of the total flux densities determined from observations of the source on 1984 May 31 and June 1. The data reduction has not yet been completed. In all three cases, the 1σ noise level in apparently blank fields of the maps near the position of the supernova is more than three times smaller.
[3] Estimated angular diameter of a uniform sphere model of the brightness distribution of the supernova. For the meaning of the errors, see the text.

To check the sensitivity of our result to the choice of antennas, we repeated the analysis with data from subsets of antennas only. The results are given in Table 3. The estimated parameters from this analysis agree with the equivalent parameters obtained from the analysis of all data to within the errors of the latter parameters, indicating that our parameter values are not sensitive to the choice of the VLBI array; the accuracies of our parameter values, however, are.

The correlated flux densities and their statistical standard deviations were divided by the estimated total flux densities, S_{tot}, for the purpose of better comparison between observations at different epochs, and then plotted together with the predictions of the models as a function of projected baseline length, $(u^2 + v^2)^{\frac{1}{2}}$, in Figure 3a-d.

Table 3
Parameter Estimates for SN 1979c at 5.0 GHz
Using Data of Subsets of Antennas Only

Date of Observation	Antennas	S_{tot}[1] (mJy)	Angular[1] Diameter (mas)
1982 Dec. 8	BGOY	$5.30 \pm .16$	$1.05 \pm .05$
	BY	$5.45 \pm .21$	$1.12 \pm .05$
	BGO	$5.05 \pm .64$	$.92 \pm .22$
1983 Dec. 1	BGOY	$4.08 \pm .14$	$1.43 \pm .05$
	BY	$4.23 \pm .19$	$1.45 \pm .05$
	BG(+Y for S_{tot})	$4.21 \pm .22$	$1.37 \pm .21$

[1] Errors denote statistical standard deviations.

Having assumed a uniform sphere model for the brightness distribution of the supernova, we obtained from the least-squares analysis a rather conservative estimate of the angular diameter. Had we assumed a ring model instead, we would have obtained an angular diameter for the source ~ 1.65 times smaller than the former, the smallest diameter allowed by the observed correlated flux densities (see, e.g., Bracewell 1978, and Marscher 1985 for more detail on the model dependence of angular diameter determinations). The values for the angular diameters of the supernova for both models are plotted in Figure 4 as a function of time following the explosion.

The 5.0 GHz data points show that, with high probability, the supernova is expanding. Assuming an indisputable lower limit of the distance to the supernova of several Mpc, we could assert this result to be the first direct observation that material is ejected from a supernova with a velocity of several 1000 km s^{-1}. Further, the two 5 GHz data points together with the zero point at the time of explosion allow the deceleration (or acceleration) parameter m, from $\theta \alpha t^m$, to be estimated. Assuming that the quoted errors for the angular diameter determinations are independent, we get $m = 1.31 \pm 0.45$.

SN 1979*c* 5.0 *GHZ* 1982 *DEC* 8

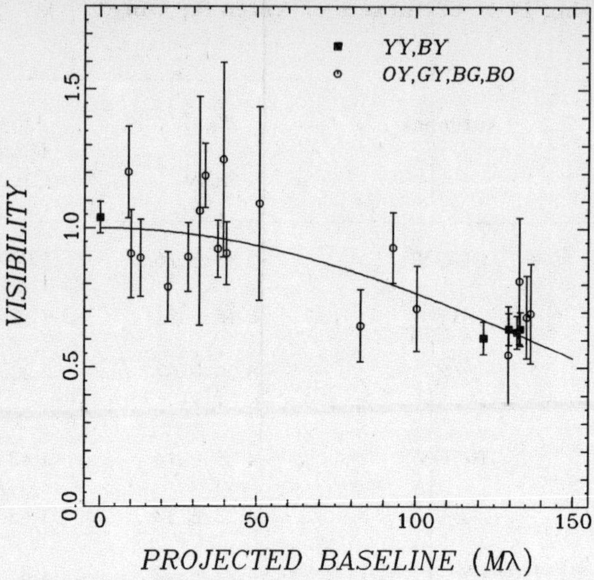

SN 1979*c* 5.0 *GHZ* 1983 *DEC* 1

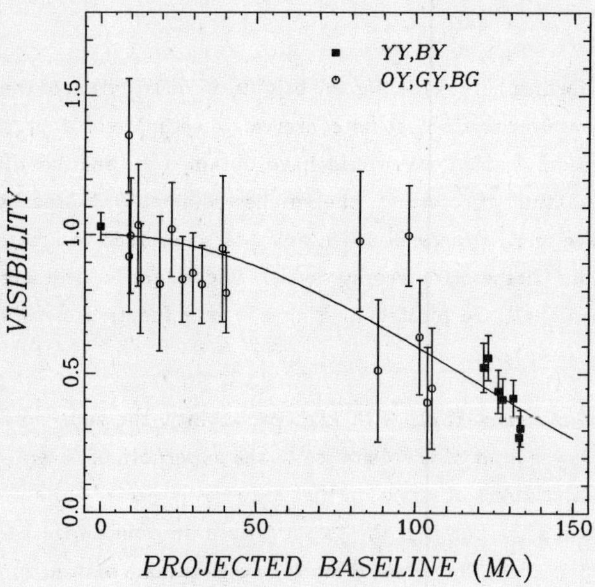

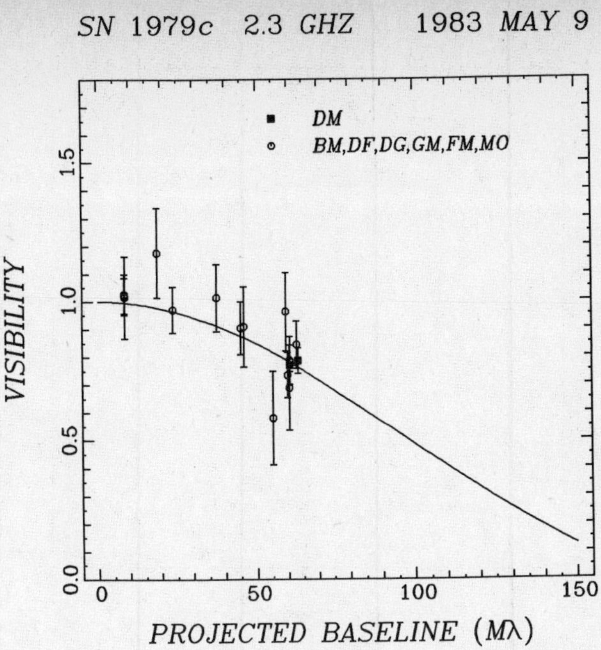

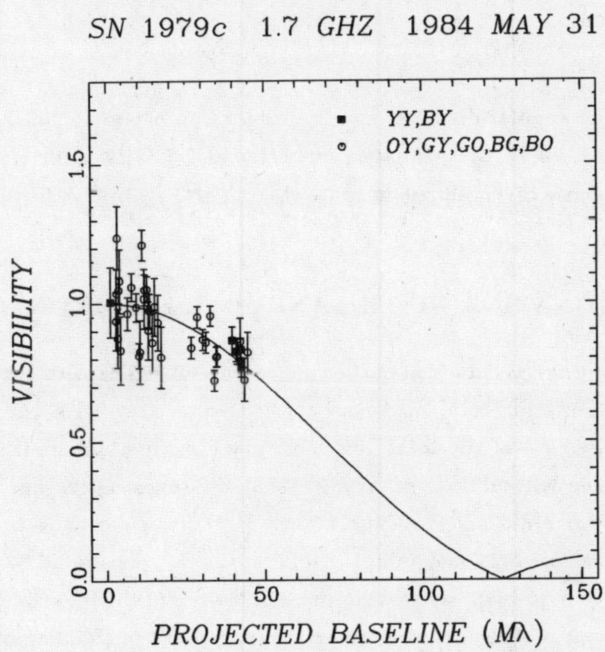

Figures 3a-d. Visibility amplitudes and predictions from the fit of a uniform sphere model. The filled squares show most significant data points. Antenna pairs are listed in order of increasing baseline length.

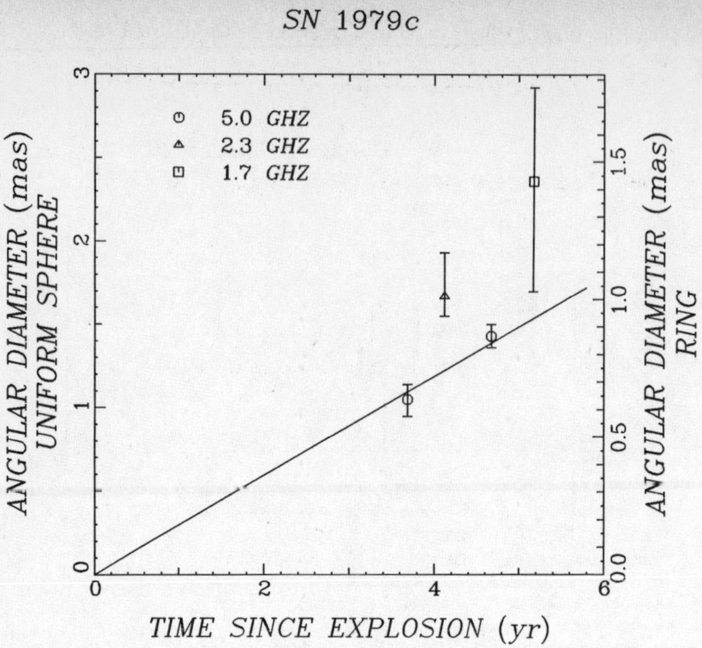

Figure 4. The angular diameter determinations assuming two (extreme) models of the brightness distribution. The solid line represents a least-squares fit to the assumed zero-point origin and the two 5.0 GHz values.

The value for the angular diameter at 2.3 GHz is clearly larger than the interpolated value for the diameter at 5.0 GHz. Unfortunately our value at 1.7 GHz is too inaccurate to describe the frequency dependence of the diameter more thoroughly.

IV. DETERMINATION OF THE ANGULAR DIAMETER OF SN 1980k

We obtained a marginal detection of the supernova at 2.3 GHz with the DM interferometer. As we could not measure the total flux density of the supernova with our telescopes, we used an approximate upper limit of the total flux density, which we obtained from measurements at 1.5 and 5.0 GHz made with the VLA (Weiler 1985) two days before our VLBI observations. Assuming that the source did not vary significantly in the two days and assuming a power law spectrum for the source, we obtained an upper limit of the flux density at 2.3 GHz on 1983 May 7 of 1.5 mJy (see inset of Figure 5). By varying the sensitivities of the telescopes as described above and taking into account a 1σ error, we derived an upper limit of the diameter, θ, for a uniform sphere model of the brightness distribution of SN 1980k (2.542 yr after 1980 Oct. 17, the assumed date of explosion) of:

$$\theta < 2.3 \text{ mas} .$$

The normalized correlated flux densities and the predictions from the model fit are shown in Figure 5.

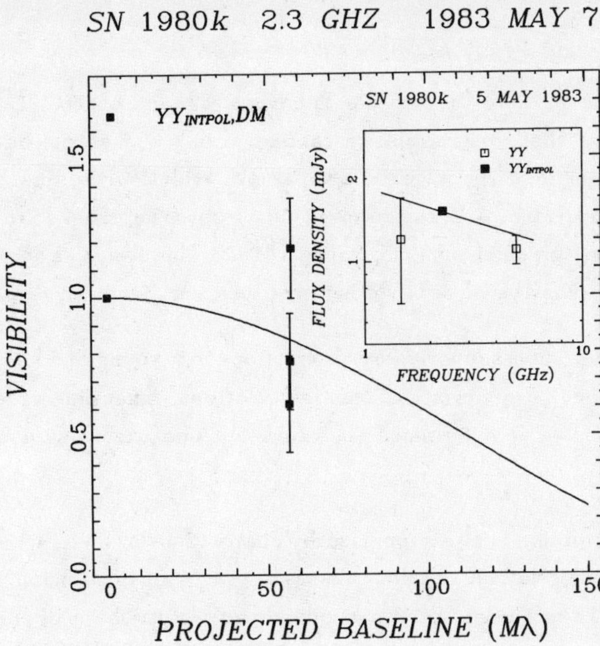

Figure 5. Visibility amplitudes of SN 1980k and the predictions from a fit of a uniform sphere model. The zero-baseline value is the upper limit obtained from interpolating between measurements made with the VLA (see inset).

V. THE DISTANCES TO M100 AND NGC 6946 AND THE HUBBLE CONSTANT

In order to estimate distances to the parent galaxies of the supernovae and to infer H_o, we want to assume that

(1) SN 1979c and SN 1980k are indeed situated in M100 and NGC 6946, respectively;

(2) the observed widths of optical spectral lines reflect expansion of the photosphere;

(3) the expansions of the photosphere and the radiosphere are isotropic;

(4) the expansion of the radiosphere with angular diameter θ can be described as $\theta \alpha t^m$, with t as the time since explosion;

(5) the expansion velocity of the radiosphere extrapolated back to a time a few months after the explosion is larger (by a frequency-dependent factor α_ν) than, or approximately equal to, the maximum expansion velocity of the photosphere.

Among these assumptions the first one is obviously the strongest. The second one is not only based on the fact that every explosion causes ejection of material but also based on the finding that at least in the early stages of a supernova evolution the time dependence of both the photospheric temperature and the received flux from a supernova are consistent with an expanding blackbody (e.g., Kirshner 1985; Panagia 1985). However, to a small extent, additional broadening of the lines may be caused by other processes, e.g., turbulence in the emission region.

The third assumption is supported by observations that young and medium aged SNRs are mostly circularly shaped. Any deviations from non-isotropic expansion, at least in the plane on the sky, could in principle be determined via VLBI, but our present data are not sufficient for such determinations.

The fourth assumption is also supported by observations of SNRs whose evolution can be described by the above relation. Again, this assumption can in principle be tested by VLBI observations and has in fact been tested partially already on the basis of our present data.

The fifth assumption is based on the fact that particles are accelerated in shock front regions and thus are likely to radiate radio waves (see, e.g., Chevalier 1982). The question is whether the shock front is the only plausible source for the observed radio radiation or whether much more luminous sources of radio radiation are situated far inside the shock front sphere and the photosphere. We know that the photospheric material, if more or less uniformly distributed, is very optically thick for radio waves. The radio waves of the sun, e.g., are emanating from outside the photosphere. Thus, in any model for which radio radiation is penetrating the photosphere, one has to assume the development of a filamentary structure for the material very soon after the explosion (Pacini and Salvati 1981). No matter how likely or unlikely such a process could be, I here only want to assume that the volume emissivity of a hypothetical radio region inside the shock front sphere is not larger than the volume emissivity of the shock front sphere itself. The limiting case in my assumption for which the volume emissivities are equal and for which the inner emission region fills the whole volume that is left by the outside emission region is described by the brightness distribution model of a uniform sphere.

With these assumptions, we can estimate distances, D, of the galaxies. The angular diameter of the radiosphere, θ, at time t after the explosion is given as $\theta = \alpha_\nu \, \theta_{ph} \left(\frac{t}{t_{ph}}\right)^m$, with θ_{ph} as the angular diameter of the photosphere at time t_{ph} and α_ν as a factor defined in the above list of assumptions. We further have $\dot{\theta}_{ph} = \frac{2v_{ph}}{D}$, with v_{ph} as the maximum velocity estimated from the width of the lines in the optical spectrum. The distance is then given as:

$$D = \frac{2 \cdot \alpha_\nu \cdot v_{ph} \cdot t}{\theta \cdot m} \left[\frac{t}{t_{ph}}\right]^{m-1} .$$

As we assume $\alpha_\nu \gtrsim 1$ and as we observed that the angular diameter of SN 1979c is smaller at the higher frequency, we further assume that the most precise estimate of the distance can be made with observations at the highest possible radio frequency. Inspecting the spectra of SN 1979c (Branch *et al.* 1981), we find a maximum velocity $v_{ph} \sim 10500$ km s^{-1} indicated by the extension of the blue wing of the $H\alpha$ profile in the spectrum on 1979 Jun. 26 (i.e., $t_{ph} = 7.43 \times 10^6$ s). With $0.86 \leq m \leq 1.76$ and with $t = 1.39 \times 10^8$ s as a time between our two 5 GHz data points, for which the angular diameter is bounded by $0.80 \leq \theta$ (mas)≤ 1.32 for the two extreme cases of brightness distributions, independent of our choice of m, we formally get for SN 1979c:

$$11.4\alpha_\nu \leq D \,[\mathrm{Mpc}] \; \leq 128\alpha_\nu; \quad \alpha_\nu \gtrsim 1 .$$

This range is considerably smaller if we allow for decelerating expansion only, i.e., $0.86 \leq m \leq 1.0$:

$$11.4\alpha_\nu \leq D \,[\mathrm{Mpc}] \; \leq 24.4\alpha_\nu .$$

Assuming that D is within ~ 10 percent equal to the distance of the Virgo center, D_o, a "local" value of the Hubble constant, H_o^{VIRGO}, can be derived by combining our distance estimates with determinations of the mean velocity of Virgo, v_o, and corrections for local streaming motions, v_{infall}:

$$H_o^{VIRGO} = (v_o + v_{infall})/D_o .$$

With $v_o = 1019 \pm 51$ km s^{-1} (Mould, Aaronson, and Huchra 1980), and $v_{infall} = 250 \pm 61$ km s^{-1} (Aaronson *et al.* 1982) and the assumption that the 10 percent error component of D_o and the errors of v_o and v_{infall} are independent, and thus do not contribute significantly to our uncertainty of H_o^{VIRGO}, our latter bounds on the distance to M100 are equivalent to:

$$110\alpha_\nu^{-1} \gtrsim H_o^{VIRGO} \,[\mathrm{km\ s^{-1}\ Mpc^{-1}}] \gtrsim 50\alpha_\nu^{-1} .$$

These bounds are ~ 7 percent smaller if we assume $v_o = 967 \pm 53$ km s^{-1} (Kraan-Korteweg 1981) and a mean of nine previous determinations of various authors of $v_{infall} = 220 \pm 50$ km s^{-1} (Tammann and Sandage 1985). With the global value of the Hubble constant, H_o, of $H_o = (50 \pm 7)(21.6/D_o)$ km s^{-1} Mpc^{-1} (Tammann and Sandage 1985; Sandage and Tammann 1985), we get:

$$100\alpha_\nu^{-1} \gtrsim H_o \,[\mathrm{km\ s^{-1}\ Mpc^{-1}}] \gtrsim 40\alpha_\nu^{-1} .$$

Comparing our estimates with those obtained by de Vaucouleurs ($D = 11.5$ Mpc for a group of galaxies in Virgo, or equivalent to $H_o \sim 100$ km s^{-1} Mpc^{-1}, 1975) and by Sandage and

Tammann ($D = 22.2$ Mpc for M100, 1979, or equivalent to $H_o \sim 50$ km s^{-1} Mpc^{-1}, 1985), we find that our estimates are consistent with those of these authors.

Had we assumed uniform expansion, i.e., $m = 1$, our lower bound for D would have been $14.8\alpha_\nu$ Mpc, and our upper bound for H_o $77\alpha_\nu^{-1}$ km s^{-1} Mpc^{-1}; both bounds would have been in disagreement with de Vaucouleurs' estimates. Clearly, a more accurate determination of m via more (and more accurate) VLBI observations would place more accurate bounds on the distance and on H_o.

From the spectrum of SN 1980k (Hempe 1983; see also Kirshner 1985), we find a maximum velocity v_{ph} of ~ 5000 km s^{-1}. If we assume uniform expansion ($m = 1$), we obtain

$$D > 2.3\alpha_\nu \text{ Mpc}; \quad \alpha_\nu \gtrsim 1 \, ,$$

which is equivalent to

$$H_o < 230\alpha_\nu^{-1} \text{ km s}^{-1} \text{ Mpc}^{-1} \, ,$$

if we couple our lower limit for the distance of NGC 6946 with Sandage's and Tammann's (1979) estimate of the distance of 10.5 Mpc and their global value of $H_o \sim 50$ km s^{-1} Mpc^{-1}. Both limits are consistent with de Vaucouleurs' and Sandage's and Tammann's estimates.

VI. CONCLUSION

VLBI observations of SN 1979c in the galaxy M100 revealed that the size of the supernova is measurable with its angular diameter being a function of frequency. From observations at two epochs, one year apart, with the same VLBI array and at the same frequency, we found that the radiosphere of SN 1979c has most likely expanded by ~ 30 percent. VLBI observations of SN 1980k in NGC 6946 allowed us to place an upper bound on the angular diameter of the supernova. The measurements place constraints on the distances to the galaxies and Hubble's constant with our lower and upper limits being consistent with de Vaucouleurs' and Sandage's and Tammann's estimates. The range of our uncertainties can probably be decreased with new and more sensitive VLBI observations of SN 1979c in the near future.

The sensitivity could be increased by recording the data on longer tapes, thus allowing longer integration times. It could also be increased by including in the array the 305 m antenna at Arecibo. The antenna's site at a low latitude would increase our northsouth resolution twofold and might enable us to place useful bounds on any elongation on the sky that SN 1979c might have.

Observations of SN 1979c at 5.0 GHz at a third epoch would probably be useful for a more accurate determination of the deceleration or acceleration of the expansion. Observations at other frequencies would allow determination of the frequency dependence of the angular diameter and perhaps allow estimation of the parameter α_ν. Finally, in the long run, further observations might enable us to determine the brightness distribution of SN 1979c. Each of these steps will be useful for placing better upper and lower bounds on the distance of M100 and on H_o.

This method of determining distances to nearby galaxies became possible through the use of the presently most sensitive interferometers, the broad-band Mark III VLBI data recording and processing system, and a data analysis scheme comparable to the "global fringe fitting" method. As about one supernova per year is detectable with our VLBI system, we hope to be able to determine distances to galaxies as far as ~ 40 Mpc and to infer the Hubble constant more accurately.

ACKNOWLEDGMENTS

I am grateful to my colleagues M. Gorenstein, C. Gwinn, J. Marcaide, and A. Rogers for their help with the observations, to I. Shapiro and K. Weiler for valuable discussions, and to L. Molnar for providing the plotting package for Figs. 3-5. This research was supported in part by the National Science Foundation under grant AST-8300796.

REFERENCES

Aaronson, M., Huchra, J., Mould, J., Schechter, P.L., and Tully, R.B. 1982, *Ap. J.* **258**, 64.

Bartel, N., Shapiro, I.I., Corey, B.E., Marcaide, J.M., Rogers, A.E.E., Whitney, A.R., Cappallo, R.J., Graham, D.A., Romney, J.D., and Preston, R.A. 1982, *Ap. J.* **262**, 556.

Bartel, N., Cappallo, R.J., Ratner, M.I., Rogers, A.E.E., Shapiro, I.I., and Whitney, A.R. 1985, *A. J.* **90**, Feb. issue.

Bracewell, R.N. 1978, *The Fourier Transform and Its Applications* (New York, McGraw-Hill).

Branch, D., Falk, S.W., McCall, M.L., Rybski, P., Uomoto, A., and Wills, B.J. 1981, *Ap. J.* **244**, 780.

Chevalier, R.A. 1982, *Ap. J.* **259**, 302.

de Vaucouleurs, G. 1975, in *Galaxies and the Universe*, eds. A. Sandage, M. Sandage, and J. Kristian (University of Chicago Press), p. 557.

Hempe, K. 1983, *Mitt. d. Astron. Gesellschaft* **60**, 107.

Kirshner, R.P. 1985, this volume.

Kraan-Korteweg, R. C. 1981, *Astr. Ap.* **104**, 280.

Marscher, A.P. 1985, this volume.

Mould, J., Aaronson, M., and Huchra, J. 1980, *Ap. J.* **238**, 458.

Pacini, F., and Salvati, M. 1981, *Ap. J. (Letters)* **245**, L107.

Panagia, N. 1985, this volume.

Rogers, A.E.E. *et al.* 1983, *Science* **219**, 51.

Sandage, A., and Tammann, G.A. 1979, *Ap. J.* **194**, 559.

Sandage, A., and Tammann, G.A. 1985, this volume.

Schwab, F.R., and Cotton, W.D. 1983, *A. J.* **88**, 688.

Tammann, G.A., and Sandage, A. 1985, preprint.

Weiler, K.W. 1985, this volume.

Weiler, K.W., Sramek, R.A., van der Hulst, J.M., and Panagia, N. 1983, in *IAU Symposium 101, Supernova Remnants and Their X-Ray Emission*, eds. J. Danziger and P. Gorenstein (Dordrecht, Reidel), p. 171.

Supernova Interaction with a Circumstellar Wind and the Distance to SN 1979c

Roger A. Chevalier
Department of Astronomy, University of Virginia

Claes Fransson
Stockholm Observatory

Abstract: There is observational evidence at a number of wavelengths for the interaction of the supernovae SN 1979c and SN 1980k with a circumstellar medium created by presupernova mass loss. Distance independent quantities can be used to develop a model for the interaction region. When these results are combined with a distance dependent quantity, the distance to the supernova is determined. The most useful distance dependent quantities are the radio angular diameter (from VLBI observations) and the X-ray flux. This method for distance determination is applied to the supernova SN 1979c.

Supernovae have recently been observed at radio, infrared, ultraviolet, and X-ray wavelengths. Many aspects of these observations can be interpreted in terms of the interaction of the expanding supernova gas with a slow moving circumstellar medium created by presupernova mass loss (see Chevalier 1984 and Fransson 1984b for reviews). If a consistent model is developed, there is the possibility of determining the distance to the supernova when quantities such as the VLBI angular diameter become available. This angular diameter has been measured for SN 1979c (Bartel et al. 1983; Bartel 1985) and we concentrate on the properties of the circumstellar interaction for this supernova.

The hydrodynamic interaction between the supernova and circumstellar gas has been described by Chevalier (1982a,b). The circumstellar medium created by a stellar wind can plausibly be described by $\rho_{cs} = \dot{M}/(4\pi v_w r^2) = Br^{-2}$, where $\dot{M}$ is the mass loss rate and v_w is the wind velocity; B is taken to be constant. For a red supergiant, the progenitor of a Type II supernova, $\dot{M}$ is typically in the range 10^{-6} to $10^{-4} M_\odot$ yr^{-1} and v_w is about 10 km s^{-1}. The time for the wind to expand to the radii of interest is generally smaller than the timescale for evolutionary changes in the stellar envelope.

A few days after the explosion, the supernova gas is approximately freely expanding, i.e. $v = r/t$. We assume that the density profile is a power law, so that $\rho_{sn} = At^{n-3}r^{-n}$. Chevalier and Jones (1984) have shown that the power law form is the profile for the outer parts of the expanding gas resulting from a variety of initial stellar configurations. Also, numerical calculations show that the profile resulting from the explosion of an actual stellar model is well fit by a small number of power law sections.

If the power law assumption is adequate, the radius of the contact discontinuity between the supernova and the circumstellar gas is

$$R_c = \left(b\,\frac{A}{B}\right)^{\frac{1}{n-2}} t^{\frac{n-3}{n-2}} \tag{1}$$

where b is a dimensionless constant. The value of b depends slightly on whether cooling is important for the shocked gas. Chevalier (1982a) gave the structure of the interaction region for the adiabatic case using similarity techniques and Chevalier (1982b) gave an analytic solution for the shell evolution in the case that both shells cool. Table 1 lists some properties for the case in which the inner shell (shocked supernova gas) cools and the outer shell does not. This case is of particular interest for SN 1979c. The Table lists the ratio of outer shock radius to inner shock radius R_1/R_2, the ratio of densities at the shock waves ρ_2/ρ_1, and the ratios of masses in the shells M_2/M_1. The subscript 2 refers to shocked supernova gas and 1 to shocked circumstellar gas.

Table 1

Properties of the Interaction Region

n	R_1/R_2	b	ρ_2/ρ_1	M_2/M_1
07	1.299	0.25	6.8	0.77
12	1.226	0.036	41	2.5
20	1.207	0.0093	156	5.2

In the present model, the radio emission is from the interaction shell between the supernova gas and the circumstellar gas. The VLBI observations give the angular radius of this shell. The most direct way to obtain the shell radius is from the velocity of the shell. We believe that the late time Hα line emission is from a dense, cool shell at the inner edge of the interaction region and the line profile leads to the gas velocity at the inner edge, v_2. If this velocity is measured at time t_1, the outer shock radius is

$$R_1(t_1) = \left(\frac{R_1}{R_2}\right) \frac{v_2 t_1}{m} \qquad (2)$$

where the factor $m = (n-3)/(n-2)$ accounts for the deceleration of the shell. If v_2 is not known, the maximum velocity in a line profile still gives a lower limit to the maximum velocity of the supernova gas, v_m. The outer shock radius is then $R_1 = v_m t(R_1/R_2)$. The time of measurement of the angular radius is taken to be t_2, which is unlikely to be the same as t_1. We have $R_1(t_2) = R_1(t_1)(t_2/t_1)^m$. Since the value of m is uncertain, t_2 and t_1 should be as close to each other as possible.

The value of m for SN 1979c is not well determined, but there are some clues. The most direct determination would be from VLBI observations over a range of times. There are VLBI observations at 5 GHz on 8 Dec. 1982 and 1 Dec. 1983 (Bartel 1985). The error limits give a range of m from 0.8 to 1.9. The present model requires $m < 1$; a large value of n is indicated. However, a 2.3 GHz VLBI measurement on 9 May 1983 gives a large angular diameter, so the uncertainties are large. Fransson (1984a) noted that the change in the maximum Hα velocity gives an estimate of m (or n) because of the deceleration of the shell. He deduced $n > 11$. Another method involves fitting the radio light curve. Chevalier (1984) deduced that $n > 12$ for SN 1979c on the basis of radio observations, but this result is model dependent. Finally, model atmosphere studies which attempt to fit the available spectrophotometry should place constraints on n. Such studies do not yet exist for SN 1979c. In our discussion, we consider two possible values of n : $n = 12$ and $n = 20$, which correspond to $m = 0.9$ and $m = 0.94$.

The Hα line profile was last observed on 10 April, 1980 by Kirshner and Chevalier (1984). The line profile can be fitted by emission from a shell with a velocity range from 7500 to 9000 km s^{-1}. This velocity range is higher than that expected for typical supernova envelope matter, so it is plausible that the emission is from the interaction shell. The observed velocity range is larger than that expected from spherically symmetric model; this could be due to the Rayleigh-Taylor instability. We choose $v_2(t_1) = 8250$

km s^{-1} with an error of ± 10%. With t_1 = 374 days we have R_1 = 3.6×10^{16} cm (n = 12) and R_1 = 3.4×10^{16} cm (n = 20).

The VLBI angular radius was measured on 8 Dec. 1982, which we designate t_2 = 3.691 yr. The angular radius was 0.523(-.049/+.044) milliarcsec (Bartel 1985). This angular radius assumes that the emitting region is a uniform sphere. For the present model, the emitting region is a shell. Marscher (1985) has calculated the effect of shell emission on the deduced angular radius and finds that for R_2/R_1 = 0.75, the shell radius is reduced by a factor of 1.25. The distance to the supernova is D = R_1/θ, where θ is the corrected angular radius in radians. From the above discussion, we find $R_1(t_2) = 1.15 \times 10^{17}$ cm (n = 12) and $R_1(t_2) = 1.13 \times 10^{17}$ cm (n = 20), which lead to D = 18.4 Mpc (n = 12) and D = 18.1 Mpc (n = 20). There is considerable uncertainty in this distance both because of the variations in the observed θ (see above) and uncertainties in the model (e.g. is the expansion really spherically symmetric?). However, there are checks on the model from all of the observational material on SN 1979c.

The radio light curves for SN 1979c show weak emission at early times (Weiler et al. 1983), which can be attributed to free-free absorption by unshocked circumstellar gas (Chevalier 1982b). From a model fit to the light curve, Chevalier (1984) deduced that the optical depth was unity at 20 cm on t = 950 days. When this information is combined with the shell radius, we find B = 3.3×10^{14} (n = 12) and B = 3.2×10^{14} (n = 20). These values assume that the temperature of the circumstellar medium is 10^4K; a higher temperature implies a higher value of B. The mass loss rates corresponding to these values of B are $\dot{M} = 6.6 \times 10^{-5} M_\odot yr^{-1}$ (n = 12) and $\dot{M} = 6.3 \times 10^{-5} M_\odot yr^{-1}$ (n = 20) for a wind velocity of 10 km s^{-1}. Once R_1, B, and n are specified, the circumstellar interaction is completely determined. The above discussion implies $\log_{10} A = 113.01$ (n = 12) and $\log_{10} A = 184.91$ (n = 20) in cgs units.

From the interaction model, it is possible to calculate the time, t_c, at which the cooling time for gas at the inner shock wave is equal to the age of the supernova. Using the cooling curve of Raymond, Cox, and Smith (1976), we find t_c = 2.9 yr (n = 12) and t_c = 100 yr (n = 20). For t < t_c, the inner shock wave is radiative and a dense, cool shell is created. The values of t_c are consistent with the interpretation of the Hα emission as being from a dense shell. It can be seen that t_c is sensitive to n. A dense shell is not formed for n = 7 at an age of 1 year.

The radiation from the cooling shock wave is primarily at X-ray wavelengths. There are limits on the X-ray flux from SN 1979c determined by the Einstein Observatory. Palumbo et al. (1981) found that L_x (0.1 - 4.5 keV) < 3.0×10^{39} (D/10 Mpc)^2 erg s^{-1} on 10 Dec., 1979 (t = 259 days). Our model predicts that on t = 259 days, the total inner shock luminosity was $L_t = 2.5 \times 10^{40}$ erg s^{-1} (n = 12) and $L_t = 1.3 \times 10^{40}$ erg s^{-1} (n = 20) and the inner shock temperature was $T_2 = 1.4 \times 10^7$ K (n = 12) and $T_2 = 3.8 \times 10^6$ K (n = 12). Most of L_t is radiated in the Einstein Observatory band, but approximately half of the luminosity is directed in toward the supernova and is absorbed. The outward directed luminosity is somewhat higher than the observed upper limit. However, this radiation is absorbed by cool gas in the dense shell and in the circumstellar medium. For n = 12 - 20, the dense inner shell is the dominant absorber. At t = 259 days, the optical depth at 1 keV is 5.2 (n = 12) and 9.7 (n = 20). The outgoing luminosity is heavily absorbed by the dense shell. After several years the absorption is greatly reduced and L_t drops slowly, so there is still the possibility of observing X-rays from SN 1979c with an X-ray telescope which has greater sensitivity than the Einstein Observatory. X-ray emission is potentially a useful distance indicator, but an X-ray light curve and temperature are probably needed to sort out the effects of absorption and various values of n.

In our model it is the absorption of X-rays by the dense shell which gives rise to the Hα line emission. The density in the shell is high (> 10^{10} cm^{-3}) which probably accounts for the large observed Hα/Hβ ratio. At t = 374 days, when the Hα line was observed, the total outward shock luminosity in the model was $L_t = 2.3 \times 10^{40}$ erg s^{-1} (n = 12) and $L_t = 1.2 \times 10^{40}$ erg s^{-1} (n = 20). On this day, the observed Hα luminosity was 1.0×10^{40} erg s^{-1} for a distance of 18.1 - 18.4 Mpc and a factor of 2 correction for absorption within the line. The implication is that the conversion of the shock luminosity to Hα luminosity is >30% efficient. When compared to calculations of similar situations for quasars (e.g. Kwan and Krolik 1981), these efficiencies appear to be higher than expected by a factor of a few. Possible resolutions of this problem are that the supernova is closer than 18 Mpc or that the densities at the shock waves are underestimated (e.g. B is actually higher than the above estimate). In any case, a detailed theory for the production of the Hα emission is needed.

Another piece of information available for SN 1979c is the density at a particular velocity in the freely expanding supernova gas. Fransson et al. (1984) studied the ultraviolet emission lines in late April, 1979 and deduced that $n_e = 4 \times 10^9$ cm^{-3} at v = 8400 km s^{-1}. We have reexamined the spectra and noted that 8400 km s^{-1} is actually only a lower limit to the velocity because of the effect of many absorption lines. The velocity could be as high as 11,000 km s^{-1}. This information determines the value of $A = \rho t^3 v^n$. For the possible range of v, we have $\log_{10} A = 67.29 - 68.11$ (n = 12) and $\log_{10} A = 183.31 - 185.65$ (n = 20). The value of A is not highly constrained, but the possible range does cover the value determined from the model.

The model must also be consistent with the highest velocity observed in the supernova gas. This velocity is limited by interaction with the circumstellar gas. Branch et al. (1981) note that the HeIλ5876 absorption line extended to a velocity of 11,700 km s^{-1} on about 25 Apr., 1979. This date is t = 24 days, for which the model predicts a maximum velocity in the supernova gas of 10,900 km s^{-1} (n = 12) and 10,200 km s^{-1} (n = 20). The model velocities are slightly less than the observed lower limit, but the values agree within the anticipated error limit.

Other observational data which relate to the circumstellar interaction are the radio flux and the infrared flux. While the observed fluxes are consistent with the model (Chevalier 1982; Dwek 1983), they are not useful as distance indicators. The generation of the relativistic electrons and magnetic field responsible for the radio synchrotron emission is not well understood. The infrared emission depends on the poorly known properties of the circumstellar dust.

In summary, a circumstellar interaction model has been developed for SN 1979c which is consistent with the available observational data. When this model is combined with the VLBI radio angular diameter, a distance to the supernova is derived. The degree of uncertainty in this distance is difficult to estimate. VLBI mapping which shows a shell of emission would give greater confidence in the model. A prediction of the model is that SN 1979c may be still detectable as an X-ray source with a sensitive X-ray telescope. The model also shows the importance of late time Hα observations of Type II supernovae.

This research was supported in part by NSF grant AST 80-19569 and by the Swedish National Science Research Council. A more complete description of the work is in preparation.

References

Bartel, N. 1985, this volume.

Bartel, N., Gorenstein, M. V., Marcaide, J. J., Rogers, A. E. E., Shapiro, I. I., and Weiler, K. W. 1983, Bull. A.A.S., 15, 954.

Branch, D., Falk, S. W., McCall, M. L., Rybski, P., Uomoto, A. K., and Wills, B. J. 1981, Ap. J., 244, 780.

Chevalier, R. A. 1982a, Ap. J., 258, 790.

Chevalier, R. A. 1982b, Ap. J., 259, 302.

Chevalier, R. A. 1984, Ann. N.Y. Acad. Sci., 422, 215.

Chevalier, R. A. and Jones, E. M. 1984, in preparation.

Dwek, E. 1983, ap. J., 274, 175.

Fransson, C. 1984a, Astr. Ap., 133, 264.

Fransson, C. 1984b, in Proc. of European Astron. Meeting, in press.

Fransson, C., Benvenuti, P., Gordon, C., Hempe, K., Palumbo, G. G. C., Panagia, N., Reimers, D., and Wamsteker, W. 1984, Astr. Ap., 132, 1.

Kirshner, R. P. and Chevalier, R. A. 1984, in preparation.

Kwan, J. and Krolik, J. H. 1981, Ap. J., 250, 478.

Marscher, A. P. 1985, this volume.

Palumbo, G. G. C., Maccacaro, T., Panagia, N., Vettolani, G., and Zamorani, G. 1981, Ap. J., 247, 484.

Raymond, J. C., Cox, D. P., and Smith, B. W. 1976, Ap. J., 204, 290.

Weiler, K. W. Sramek, R. A., van der Hulst, J. M., and Panagia, N. 1983, in IAU Symp. No. 101, Supernova Remnants and Their X-Ray Emission, ed. J. Danziger and P. Gorenstein (Dordrecht: Reidel) p. 171.

MODEL AND GEOMETRY DEPENDENCE OF RADIO DISTANCE DETERMINATIONS OF EXTRAGALACTIC SUPERNOVAE

Alan P. Marscher

Department of Astronomy, Boston University

1. INTRODUCTION

Now that supernovae are being detected in the radio (see the papers by Sramek 1985 and Weiler 1985), it is possible to use angular size and velocity measurements to obtain the distances to their respective galaxies. Bartel (1985) has done just that by obtaining angular diameters from VLBI (very-long baseline interferometry) observations and expansion velocities from optical spectral line data. The application of these measurements to distance determinations requires assumptions, which lead to uncertainties in the derived distances. Other papers in this volume deal with the optical-radio velocity correspondence; here I concentrate on the uncertainties involved in extracting angular diameters from VLBI data.

At this point, the weakness of the flux densities of observed radio supernovae combines with the small number of highly sensitive antennas available to limit the information which one can extract using VLBI. This situation should improve as receiver technology advances and when (if?) the VLBA becomes operational. For the time being, we must make the simplifying assumption of spherical symmetry. One expects this theoretically, since supernovae are extremely energetic, essentially point explosions, but one can imagine that nature could be more complex. For example, if the pre-supernova wind in Chevalier's (1982, 1985) model is not spherically symmetric, neither will be the radio emission. In any case, better coverage of the spatial frequencies (*viz.*, more sensitive dishes) should uncover asymmetries if they exist.

Within the assumption of spherical symmetry, several models are possible. The pulsar-driven model of Bandiera, Pacini, and Salvati (1984) leads to a uniform sphere of emission. This requires that the huge mass of ejected thermal gas break up into filaments shortly after the explosion, a condition argued against by Chevalier (1982). Chevalier's model involves a shock in an outer shell where the ejecta interact with a pre-supernova wind. This produces a more-or-less uniform synchrotron shell of thickness xR, where R is the outer radius of the ejecta, with $x \approx 0.07$ to 0.25. Chevalier favors external thermal free-free absorption by the wind as the cause of the spectral turnover and sharp turn-on in the radio. Although his model fits the radio light curves fairly well, other models might be possible. For example, Razin-Tsytovich suppression within the shock cannot be excluded as the mechanism responsible for the spectral turnover and rapid rise in flux density. Also, a uniform shell which is self-absorbed by free-free opacity is a possibility, although the currently existing model (Marscher and Brown 1978) is *ad hoc* and predicts a slightly less rapid turn-on than is observed.

Bartel (1985) interprets his VLBI observations by fitting the brightness distribution of a uniform, optically thin sphere to the data. He adjusts the angular diameter of the model sphere until the best fit is obtained. If the correct model is not a uniform sphere, the angular diameter thus obtained is an overestimate, which leads to an underestimate of the distance and an overestimate of H_0. I show below how the ratio of inferred to true angular diameter, and therefore the derived value of H_0, depends on the geometry of the source for several possible models.

II. INTERFEROMETRIC VISIBILITY CURVES OF MODEL SOURCES

An interferometer does not measure directly the brightness distribution of the source. Rather, the VLBI data represent components of the Fourier transform of the brightness distribution. The normalized amplitudes of the Fourier components are collectively called the visibility curve of the source. The phases are not in general measurable in VLBI owing to ionospheric propagation effects. However, the "closure phase" of a closed loop of baselines (e.g., $\phi_{12} + \phi_{23} - \phi_{13}$) is independent of these effects and can be used to determine whether there are any anisotropies in the brightness distribution. For a symmetric source, the closure phase is either 0^0 or 180^0.

The visibility curve of a source which is circularly symmetric on the sky is given by (e.g., Bracewell 1965)

$$V(b) = \int_0^1 I_\nu(\rho) J_o(\pi b \rho \theta) \rho d\rho,$$

where I_ν is the intensity distribution [normalized such that $V(0) = 1$], ρ is the normalized angular distance from the center of the source, θ is the angular diameter of the source in radians, b is the baseline (separation of the telescope pair, in wavelengths), and J_o is the Bessel function. For a uniform disk, one obtains $V(b) = 2J_1(\pi b\theta)/(\pi b\theta)$; for a uniform sphere $V(b) = 3\sqrt{\pi/2}\ J_{3/2}(\pi b\theta)/(\pi b\theta)^{3/2}$; and for a very thin ring $V(b) = J_o(\pi b\theta)$. More complicated models must be integrated numerically.

Figure 1 shows the intensity distributions at different frequencies for three source models: (a) uniform, synchrotron self-absorbed sphere (Bandiera *et al.*, 1984); (b) uniform, free-free self-absorbed shell with $\Delta R = 0.25R$; and (c) uniform shell, $\Delta R = 0.25R$, externally absorbed by free-free opacity in a surrounding wind (Chevalier 1982, 1985). The frequency ν_m corresponds to the spectral turnover. Figure 1 (d) shows the effects of varying the shell thickness at optically thin frequencies for models (b) and (c). The self-absorbed sphere and shell intensity profiles become identical to that of a uniform disk at highly opaque frequencies, $\nu \ll \nu_m$.

Since the intensity profiles are different for the various models, so should be the visibility curves. Figure 2 illustrates this by comparing the visibility curves for four possible source geometries (with the ring being an unlikely candidate but useful as a limiting case). As a general rule, the models whose brightness distributions are more concentrated toward the outer part of the source have visibility curves which drop off more rapidly as the baseline b increases. Therefore, if one assumes wrongly that they are optically thin, uniform spheres, the derived value of the angular diameter θ will be overestimated. A convenient parameter for comparison is the baseline for which $V = 0.5$, $b_{1/2}$. The value of θ inferred by a VLBI observer who assumes an optically thin, uniform sphere is inversely proportional to $b_{1/2}$. From Figure 2, we find that $\theta_{inferred} = 1.12\ \theta_{true}$ for a uniform disk and $\theta_{inferred} = 1.65\ \theta_{true}$ for a very thin ring. An optically thin spherical shell has a brightness distribution which, for small values of $\Delta R/R$, can be approximated as that of a disk of radius R plus that of a ring of radius $R - (\Delta R/2)$ (cf. Fig. 1d). This is a good description for $x = \Delta R/R \lesssim 0.2$, but breaks down for thicker shells. At optically thin frequencies, we obtain $\theta_{inferred} = 1.24\ \theta_{true}$ for $x = 0.25$, $\theta_{inferred} = 1.27\ \theta_{true}$ for $x = 0.2$, $\theta_{inferred} = 1.315\ \theta_{true}$ for $x = 0.1$, and $\theta_{inferred} = 1.34\ \theta_{true}$ for $x \lesssim 0.05$.

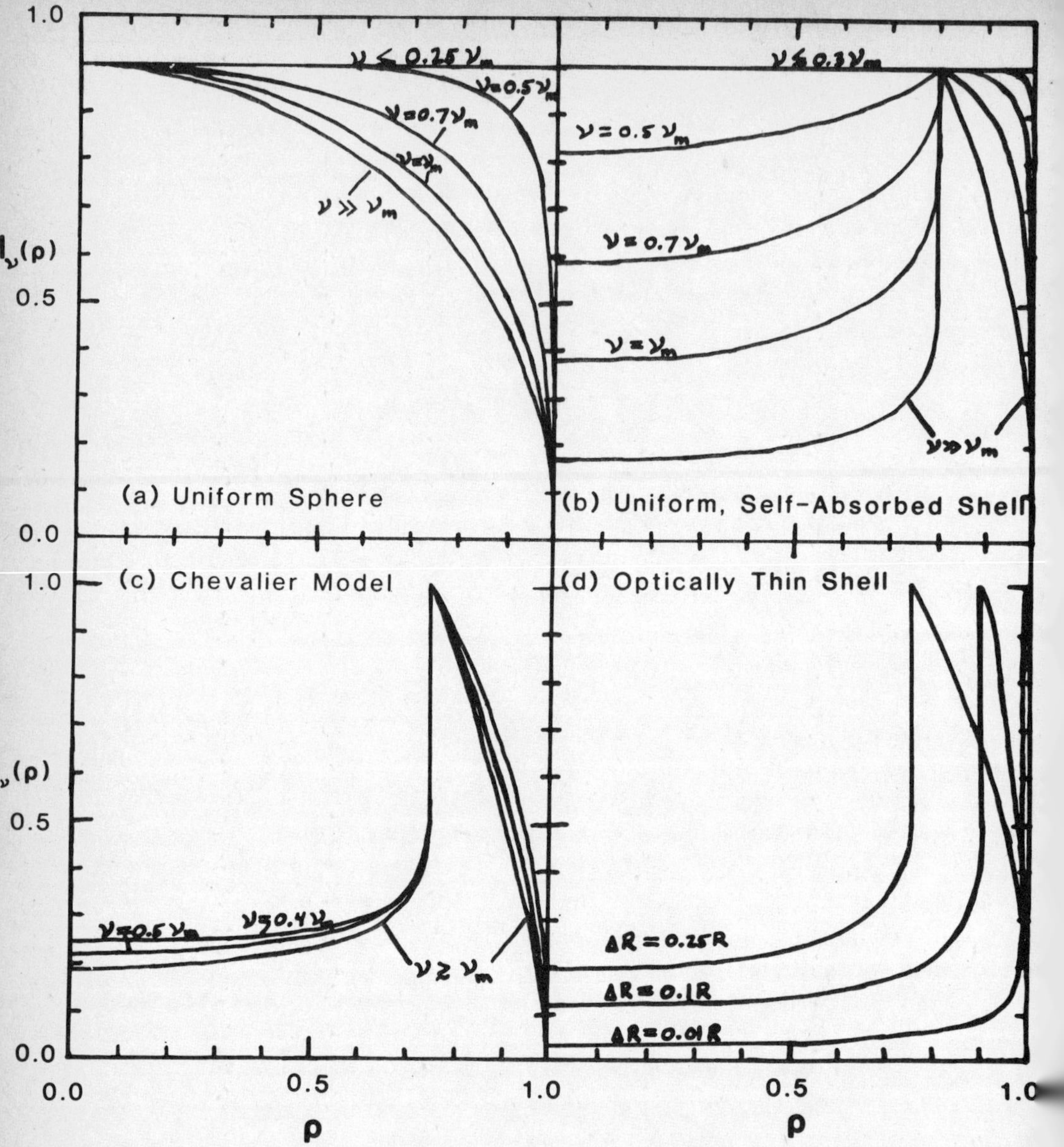

Fig. 1. Intensity profiles for various models.

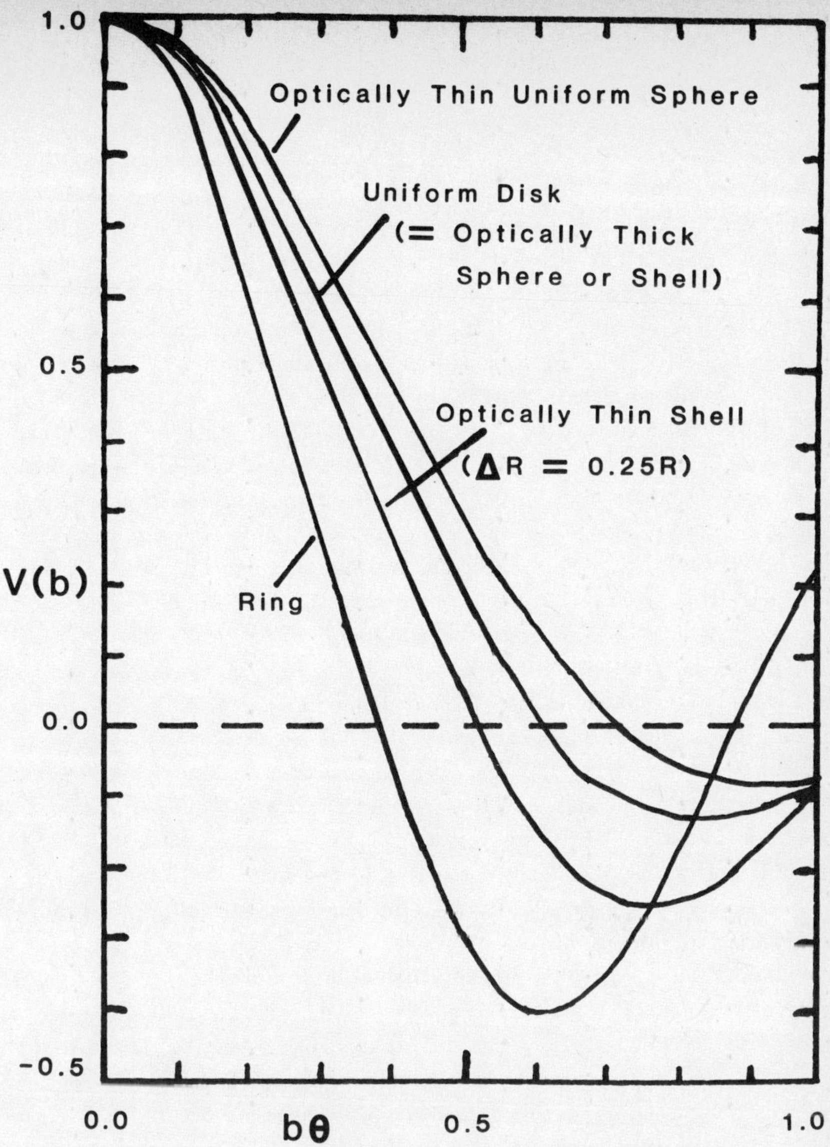

Fig. 2. Visibility Curves for Various Models.

Units of bθ are 413 Million wavelengths-milliarcsec

III. OBSERVATIONAL DETERMINATION OF SOURCE GEOMETRY

The previous section describes the current state of affairs. Since it is not certain which geometry is correct, a systematic uncertainty in the derived value of H_0 must be included in addition to the observational errors. If one is willing to consider the extreme case of a ring as a possible geometry, the value of H_0 obtained by assuming a uniform sphere could be high by as much as 65%. Bartel has taken this into account by determining θ_{true} for the two most extreme geometries in order to obtain an upper and lower bound on H_0. In the more likely case of a shell geometry (Chevalier's model does, after all, provide a good fit to the radio light curves), the overestimate is 25-34%, depending on the value of the shell thickness, $x = \Delta R/R$.

It would be highly desirable to eliminate this source of uncertainty by deciding observationally which geometry is correct. This is possible, although difficult. Figure 3 gives the visibility curves for a uniform sphere, uniform shell (nearly the same result is obtained for $x = 0.25$ as for $x = 0.05$), and thin ring, all with the angular diameters adjusted such that the curves have the same value of $b_{1/2}$. Since the experimental errors on V are typically 5-10%, we find that the geometries produce essentially the same visibility curve until V approaches zero. Since each model considered has a sharp edge, the visibility curve performs a damped oscillation about zero as b is increased. One actually observes $| V | S_\nu$, where S_ν is the total flux density. The observed visibility curve would then have a secondary maximum at long baselines, the amplitude of which depends rather sensitively on the geometry. For a ring, $| V |$ (2nd max.) $= 0.4$, while the corresponding value for a shell and sphere are 0.25 and 0.1, respectively. Unfortunately, the secondary maximum can only be observed for large values of $b\theta$. The Earth's diameter limits b, while θ becomes large as the source ages, and, in general, the flux density declines. Nature seems to fight against us here, but a slowly developing, bright, nearby radio supernova (peak flux density at 5 GHz occurring after several years, $S_\nu > 10\ mJy$ at peak, $d \lesssim 10\ Mpc$) might allow such an observation to be made, especially if improvements in VLBI arrays continue.

Another difference among the possible models is the dependence of the inferred angular size on frequency. As one observes at progressively lower frequencies, the brightness distribution changes as opacity effects become important (cf. Figure 1). Figure 4 illustrates how the inferred angular size (compared to the true angular diameter) depends on frequency for representative source models. Chevalier's model shows little frequency dependence until $\nu \lesssim 0.5\nu_m$, below which point the emission is limb-darkened and sharply cut off in flux density owing to the external absorption. The inferred angular size of a self-absorbed shell decreases toward lower frequencies as the central region of the brightness distribution fills in. The opposite behavior occurs for a sphere since the relative brightness increases in the outer regions as the source becomes opaque. Again, however, nature fights against our observing this effect. Earth-based interferometers have a maximum value of the baseline b (measured in wavelengths) which decreases linearly with decreasing frequency. When the turnover occurs at a typical radio frequency, the angular size is usually too small to resolve at that frequency. Again, a nearby radio supernova would help; otherwise a sensitive dish or two in space could give the long baselines required.

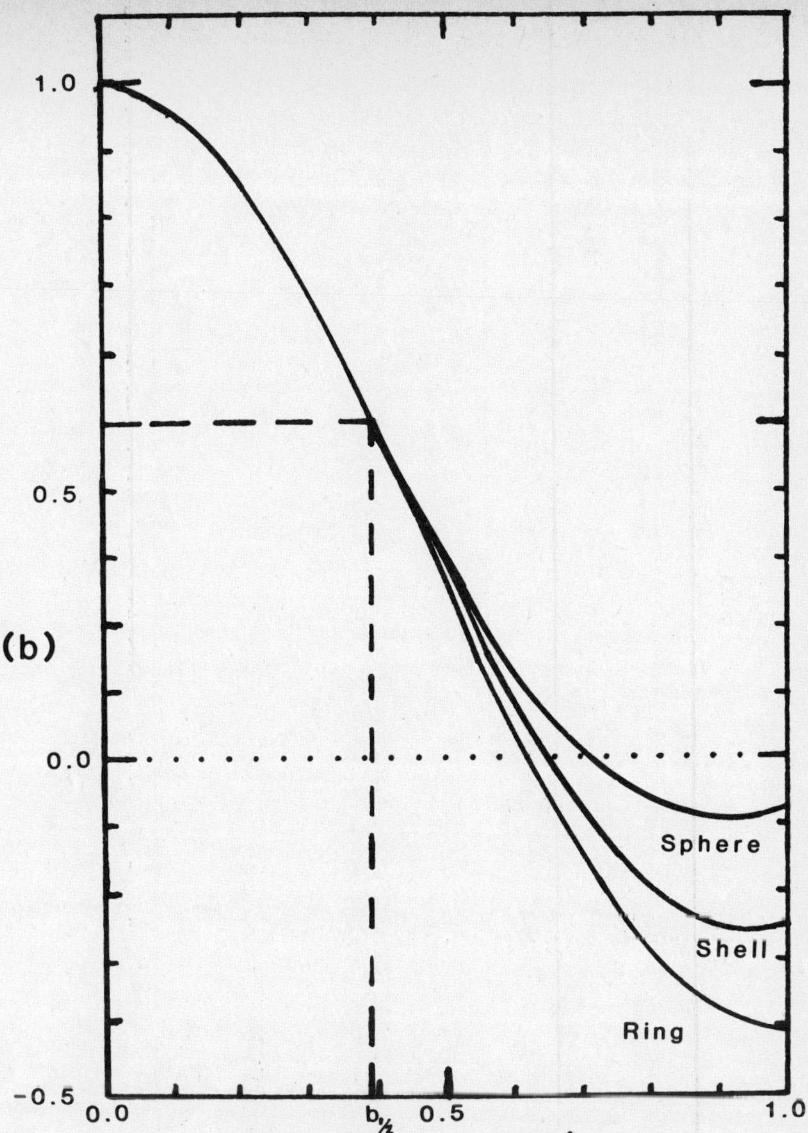

Fig. 3. Visibility Curves for Various Models with Identical $b_{1/2}$.

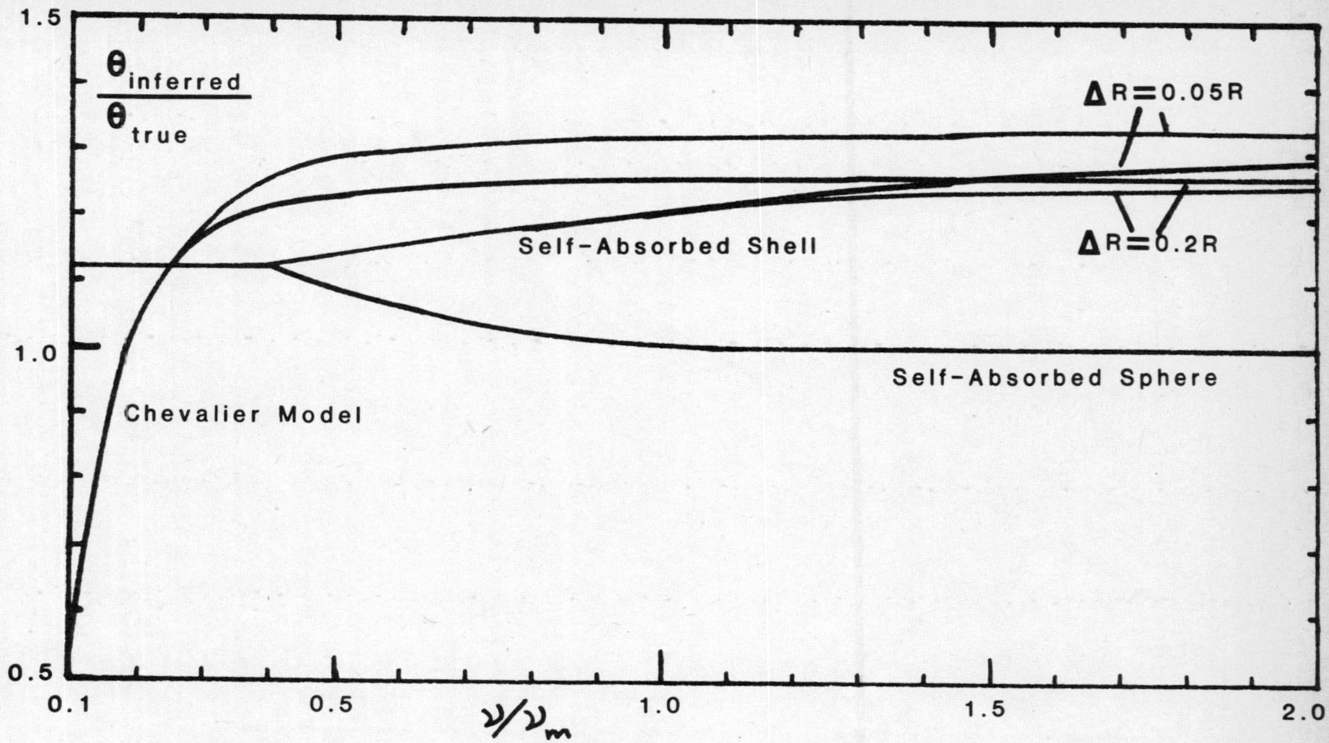

Fig. 4. Inferred Angular Diameter Relative to True Diameter if Uniform, Optically Thin Sphere Assumed.

IV. SUMMARY

I hope that the above discussion elucidates the systematic model and geometry dependent uncertainties in using radio supernovae as distance indicators. Of all the models considered, a uniform sphere is the most "conservative" in that it yields the highest value of H_0. Since Chevalier's model currently seems to best satisfy the observational and theoretical constraints, it seems likely that the value of H_0 obtained by adopting a uniform spherical geometry is high by 25-34%. However, supernovae may have a few surprises in store for us. The multifrequency VLBI data of Bartel suggest a possible frequency dependence of $\theta_{inferred}$ even though $\nu \gg \nu_m$. This should not be possible unless gradients in magnetic field and particle energy exist in the emitting region. For example, the electrons whose radiation peaks at 2.3 GHz could occupy a larger volume than those whose radiation peaks near 5 GHz. It is difficult for this to yield significantly different observed sizes, however, since synchrotron radiation is broadband, which precludes sharp frequency cutoffs within a given volume of emission. Nevertheless, we should remain aware that unexpected complicating effects such as gradients in magnetic field and relativistic particle densities and energies, scattering, coherent emission mechanisms, and inhomogeneous opacity effects could plague our efforts to use radio supernovae as distance indicators.

I thank Norbert Bartel for discussing his results prior to the Workshop. This research was supported in part by the National Science Foundation under grant AST-8315556.

V. REFERENCES

Bandiera, R., Pacini, F., and Salvati, M. 1984, *Ap.J.*, **285**, 134.
Bartel, N., 1985, this volume.
Bracewell, R.M. 1965, *The Fourier Transform and Its Applications* (New York: McGraw-Hill).
Chevalier, R.A. 1982, *Ap.J.*, **259**, 302.
Chevalier, R.A. 1985, this volume.
Marscher, A.P., and Brown, R.L. 1978, *Ap.J.*, **220**, 474.
Sramek, R.A. 1985, this volume.
Weiler, K.W. 1985, this volume.

OPTICAL SUPERNOVAE AND THE HUBBLE CONSTANT

David Branch

Department of Physics and Astronomy
University of Oklahoma, Norman, OK 73019
and
Lick Observatory, Board of Studies in Astronomy and Astrophysics
University of California, Santa Cruz, CA 95064

Abstract

Three ways to use optical supernovae to estimate the value of the Hubble constant are reviewed: (1) Type I supernovae as standard candles, (2) the Baade-Wesselink method applied to Type II and Type I supernovae, and (3) the ^{56}Ni-radioactivity method for SNe I. It is likely that $H_o \approx 50 - 60$ km s^{-1} Mpc^{-1}. If "ordinary" Type I supernovae are completely-disrupting carbon-oxygen white dwarfs near the Chandrasekhar mass, the ^{56}Ni-method provides a rigorous lower limit, $H_o \gtrsim 40$, and a firm upper limit, $H_o \lesssim 70$ km s^{-1} Mpc^{-1}.

I. Introduction

This contribution is concerned with three ways to use the optical properties of supernovae to estimate the value of the Hubble constant. The methods are discussed in order of the increasing physical understanding required for their application. The first approach, which treats Type I supernovae (SNe I) as "standard candles", is a purely astronomical one requiring no physical understanding at all. It is akin to the classical "distance ladder" approach to H_o in that it requires the distance to one ordinary SNe I to be determined by other means. The second approach, the Baade-Wesselink method, is an astrophysical one which is independent of all other distance determinations. It can be applied to the early, optically-thick phases of any supernova, and it does not require knowledge of the supernova progenitor or its explosion mechanism. The third method, based on ^{56}Ni radioactivity in SNe I, also is independent of other astronomical distances; it does, however, require the SN I progenitor to be known and its explosion mechanism to be understood.

II. Type I Supernovae as Standard Candles

Observationally, Type II supernovae do not appear to be good standard candles [1]. From the theoretical point of view, it would be surprising if they were. The appearance of SNe II almost exclusively in the arms of spiral galaxies establishes that their progenitors are massive stars, having initial main-sequence masses greater than, say, 8 $M_\odot$. Stellar evolution theory predicts that these stars develop heavy-element cores which collapse to form neutron stars or black holes. A small fraction of the gravitational energy released by the collapse goes (somehow) into a shock wave which heats and ejects the star's envelope, and the diffusive release of internal energy in the expanding envelope provides for the early SN II light curve [2-4]. The peak luminosity depends on the mass and radius of the envelope at the time of the explosion, and the amount of energy deposited in it [5]. All three of these quantities will vary from one SN II to another, not only because the progenitors come from a range of initial masses, but because even stars of the same initial mass will have a range of envelope masses and radii at the time of explosion, owing to mass transfer processes involving binary companions [6].

Type I supernovae, on the other hand, may be good standard candles. Apart from a small fraction (~ 0.1) which are noticeably peculiar spectroscopically or photometrically, SNe I show a remarkable homogeneity in light-curve shape and spectral evolution. It is not quite so straightforward to show that their peak brightnesses also are similar, but Tammann [1] (see also Sandage [7]) finds that a sample of 17 SNe I in (dust-free) elliptical galaxies satisfies

$$< M_B >= -18.19\ (\pm 0.14) + 5 \log \frac{H_o}{100} \tag{1}$$

with a dispersion about the mean of 0.58 *mag*. To ensure that departures from pure Hubble flow in the local supercluster do not affect this relationship, Arnett, Branch, and Wheeler [8] restrict the sample to six SNe I in elliptical (and lenticular) galaxies having radial velocities in excess of 3000 km s^{-1}, and find a similar result:

$$< M_B >= -18.3\ (\pm 0.2) + 5 \log \frac{H_o}{100}, \tag{2}$$

with a dispersion of 0.3 mag. Theoretically, the homogeneity is in accord with the presently popular view that the progenitors of SNe I are mass-accreting white dwarfs which undergo thermonuclear explosions upon approaching the Chandrasekhar mass.

SNe I in spiral and irregular galaxies are systematically fainter and show a larger scatter, but the brightest of them are about as bright as those in ellipticals. It is possible that this simply reflects the varying amounts of interstellar extinction of these supernovae in their generally dusty parent galaxies [1]. However, there are some reasons to suspect that SNe I in spirals have a real spread in their brightnesses. The peak absolute magnitude and the initial light-curve decay rate appear to be correlated, in the sense that the brighter the supernova the slower the decay [9-11]. Also, SNe I beyond the local supercluster appear to decay slower, on the average, than the nearer ones [11], which makes sense in terms of an observational selection effect: among remote supernovae, the discovery of the brightest ones is strongly favored. A recent spectroscopically-peculiar SN I, 1983n in M83, was undoubtedly fainter than ordinary SNe I [12], but it is not clear that this supernova has any bearing on the properties of the ordinary ones. Its initial light-curve decay rate was not particularly rapid. At present, it is hard to decide whether SNe I in spirals consist of (1) a homogeneous group of "ordinary" SNe I, plus a few peculiar ones, or (2) a single family of supernovae whose properties cover a continuous range, with the most extreme being labelled "peculiar". For the time being, it would be best to treat only SNe I in elliptical galaxies as standard candles.

It would be desirable, then, to determine the distance to one SN I in an elliptical galaxy. Unfortunately, none of the relatively nearby SNe I observed so far were in typical, dust-free ellipticals. If the standard candle approach is to be used *now*, a supernova in a spiral, irregular, or peculiar elliptical galaxy must be used as a calibrator. One can hope to adequately estimate the extinction in the parent galaxy, but it must be kept in mind that the calibrator may be *intrinsically* fainter than the remote supernovae used to establish *eqns.* (1) and (2), and that the derived value of H_o may be too high.

A list of possible calibrators is given in Table 1. The apparent blue magnitudes, B (or, in the case of the two historical Galactic supernovae, the visual magnitudes), give a rough indication of the relative distances of these events. Their suitability as calibrators is discussed briefly in the following paragraphs.

Table 1

Possible SN I Calibrators

SN	*Galaxy*	*B*	*Comment*
1572 (Tycho's)	ours	-4(V)	M_B and type uncertain.
1604 (Kepler's)	ours	-4(V)	M_B and type uncertain.
1885a	M31	6	Peculiar light curve.
1895b, 1972e	NGC 5253	8	Ordinary SNe I.
1983n	M83	11	Peculiar spectrum.
1937c	IC 4182	8	Ordinary SN I.
1954a	NGC 4214	9	Peculiar spectrum.

a) Tycho's and Kepler's Supernovae

The absolute magnitudes of these supernovae depend on their apparent magnitudes, interstellar extinctions, and distances, all of which are uncertain. de Vaucouleurs [13], however, has derived $H_o = 97 \pm 20$ km s^{-1} Mpc^{-1} on the basis of Tycho. A recent study of supernova light curves has shown that even the types are uncertain. Light curves constructed from contemporary descriptions of Tycho's and Kepler's supernovae have been regarded as consistent with the light curves of SNe I but not with those of SNe II [14-18]. But the light curves of SNe II can be subdivided into "plateau" and "linear" varieties, with the latter comprising about one third of all observed SNe II [19]. Doggett [20] now has shown that the mean light curve of *linear* SNe II is more like that of SNe I than like plateau SNe II, and that the Tycho and Kepler light curves are as much like linear SNe II as like SNe I. The properties and positions of the Tycho and Kepler supernova *remnants* may provide reasons to suspect that they were SNe I, but the most convincing part of the argument no longer seems to be valid.

b) SN 1885a in M31

The appearance of this supernova near the nucleus of the Andromeda galaxy a century ago marked the beginning of research on extragalactic supernovae. A recent thorough study of the observations of SN 1885a establishes that the light curve was much faster to decline than that of ordinary SNe I [21]. The possibility that photometric error is responsible appears to be excluded. If the justification for treating SNe I as standard candles rests on the observed homogeneity, peculiar events such as SN 1885a should not be used as calibrators.

c) SNe 1895b and 1972e in NGC 5253

Neither of these supernovae are known to have been peculiar in any respect. SN 1895b was not well observed by modern standards, but its light curve [22] and its one published spectrum [23] appear to have been normal. SN 1972e *was* very well observed [24], and is one of the two modern prototypical SNe I (the other being SN 1981b in NGC 4536 [25]). The peak apparent magnitudes of SNe 1895b and 1972e were the same to within a few tenths of a magnitude, and the outlying position of SN 1972e in the parent galaxy suggests that its internal extinction was small.

d) SN 1983n in M83

The apparent magnitude of this supernova is misleading as an indicator of distance, because it was subluminous [12]. The *shape* of the light curve was within the normal range, perhaps somewhat slower than average. The spectrum was strikingly peculiar, lacking some of the characteristic SN I features as well as the usual redward shifts of certain other features during the first month after maximum light. SN 1983n certainly should not be used as a calibrator.

e) SN 1937c in IC 4182

SN 1937c, the original prototypical SN I, was well observed both photometrically [26] and spectroscopically [27]. It had no known peculiarity.

f) SN 1954a in NGC 4214

The shape of the light curve of SN 1954a [28] was normal, perhaps somewhat fast to decay. Its spectrum [29] was peculiar. The spectral lines were unusually narrow, and extra features, not normally seen in SNe I, were present in the red.

It is interesting that although only about ten per cent of observed SNe I have been noticeably peculiar, three of the nearest six extragalactic SNe I have just been excluded on the grounds of peculiarity. The high incidence of individuality among the nearby SNe I probably should not be attributed to their being better observed than the more remote ones; the peculiarities of SNe 1885a, 1954a, and 1983n would be easily detectable at larger distances. The explanation is more likely to be that "peculiar" SNe I tend to be subluminous (recall that SN 1983n certainly was; SN 1885a also was, if H_o is much less than 100). Again, at large distances the observational selection effects favor the discovery of the brighter, "ordinary", SNe I.

Only NGC 5253 and IC 4182 emerge as parent galaxies of suitable SN I calibrators. All three of the supernovae in these two galaxies had apparent magnitudes, corrected for extinction, near $B = 8$. It follows from *eqn.* (1) or (2) that distances of roughly 2 Mpc would correspond to $H_o \approx 100$, and 4 Mpc to $H_o \approx 50$.

NGC 5253 has been classified as peculiar elliptical [30], peculiar *I*0 [31], and "amorphous" [32]. It has been thought to be a member of the Centaurus group of galaxies [33] and a companion to M83 (NGC 5236). Estimates of the distance to M83 range from $\lesssim 4$ [34] to $\gtrsim 8$ [35] Mpc. To reconcile the supernovae in NGC 5253 with $H_o \approx 100$, de Vaucouleurs [34] has suggested that NGC 5253 may be in the foreground of the Centaurus group, near 2 Mpc. Even at 4 Mpc, NGC 5253 would be a relatively small galaxy, and it is interesting that it has managed to produce two supernovae in the last century. Van den Bergh [36] has suggested that the high supernova rate in NGC 5253 and the warped distribution of neutral hydrogen in M83 may be consequences of a close tidal encounter between the two galaxies some 10^8-10^9 years ago. If NGC 5253 is only at 2 Mpc, its supernova rate per unit galaxy luminosity goes up by a factor of 4, while van den Bergh's explanation for it becomes untenable.

IC 4182, a late-type spiral or irregular galaxy, has been thought to be a member of the Canes Venatici I group [33]. Sandage and Tammann [37] (see also Sandage [7]) have determined a distance of 4.4 Mpc for IC 4182 by identifying its three brightest red stars and adopting a mean absolute visual magnitude for the three of −7.7 *mag.*, established from the brightest red stars in nearer galaxies whose distances have been determined *via* cepheids. With this distance, together with a similarly determined distance to NGC 4214, Sandage and Tammann use SNe I to derive $H_o = 50 \pm 7$ km s^{-1} Mpc^{-1}. Since SN 1954a in NGC 4214 was peculiar, this result for H_o rests primarily on the distance to IC 4182. de Vaucouleurs has not determined a distance to IC 4182 itself, but his distance to the C Vn I group of 4.9 Mpc [38] is even somewhat larger (!) than that of Sandage and Tammann. To reconcile SN 1937c with $H_o \approx 100$, IC 4182, like NGC 5253, would need to be placed in the foreground of the group to which it has been assigned.

At present, the use of SNe I as standard candles points to a low value of H_o, nearer to 50 than to 100 km s^{-1} Mpc^{-1}. The principle evidence is the Sandage-Tammann distance to IC 4182. Some weak supporting evidence may be provided by NGC 5253. It would be very important to establish an accurate, uncontested distance to IC 4182 using cepheids, perhaps with the help of the Space Telescope.

III. The Baade-Wesselink Method

As applied to supernovae, this approach is similar in spirit to the classic Baade-Wesselink method for pulsating variable stars [39,40]. The physical situations are so different, however, that each application, to supernovae and to variable stars, must be judged on its own merits. The goal is to use observations of a supernova together with simple physical principles to infer the distance of the supernova directly. In principle the method is applicable at all times when the supernova is optically thick, but in practice it is applied from maximum light to about a month later, while the photosphere is expanding.

The gist of the method is to derive the distance by matching a distance- independent "photometric" angular radius of the photosphere to a distance- dependent "spectroscopic" angular radius. The photometric angular radius is given by

$$\theta = (\frac{f_\nu}{F_\nu})^{1/2} \tag{3}$$

where f_ν is the observed monochromatic flux, corrected for extinction, and F_ν is the intrinsic flux. In practice the intrinsic flux is approximated to be that of a blackbody having a temperature equal to the optical color temperature of the supernova. For both SNe I [25] and SNe II [41], the angular radius is found to increase until about a month after maximum light, and then to decrease. The spectroscopic angular radius at time t is given by

$$\theta = \frac{R_o + V(t - t_o)}{D} \tag{4}$$

where R_o is the initial radius, V is the velocity of material at the photosphere, t_o is the time of the explosion, and D is the distance. The velocity is inferred from the spectral line profiles to an accuracy of 10 or 20 per cent, and in practice R_o is neglected. The approximation being made in *eqn.* (4), that each bit of matter has been moving at constant velocity since the time of the explosion, is expected to be an excellent one. The fact that the photosphere itself does not move at constant velocity is not relevant.

For SNe II, t_o is not known a *priori*, so observations made at several different times are used to infer t_o and D simultaneously. Kirshner and Kwan [42] derived distances of 6 ± 3 Mpc for SN 1970g in M101 and 12 ± 3 Mpc for SN 1969l in NGC 1058. These distances correspond to Hubble ratios of about 60 km s^{-1} Mpc^{-1}, but both galaxies are within the local supercluster, which has a retarded expansion rate, so the implied global value of H_o would be somewhat higher. Branch *et al.* [41] derived a distance of 23 ± 3 Mpc (internal errors) to SN 1979c in M100, in the Virgo cluster. This distance gives a Hubble ratio for Virgo of 40 km s^{-1} Mpc^{-1}, but adopting 300 km s^{-1} for our peculiar motion relative to Virgo yields a global value of H_o of about 60 ± 10 km s^{-1} Mpc^{-1}.

SNe I can be treated in the same way, or, as suggested by Arnett [43], the rise time $t - t_o$ can be estimated from observation, in which case the distance follows just by matching *eqns.* (3) and (4) at a single time. Composite pre-maximum light curves for SNe I [44,10] indicate, with only a minimal extrapolation, that the typical rise time is about 15 days. The distance to the supernova fixes its absolute magnitude, which then is used with *eqn.* (1) or (2) to give H_o. Applications to SNe I have led to low values of the Hubble constant, $\lesssim$50 km s^{-1} Mpc^{-1} [45,25].

Observationally, it is well within current capabilities to apply the Baade- Wesselink approach to supernovae beyond the local supercluster, so that correcting for local departures from pure Hubble flow, or using the standard- candle relationship between M_B and H_o for SNe I, could be dispensed with. The main reason that the necessary observations of such distant supernovae have not yet been made is that the dominant present uncertainty lies elsewhere – in the approximation of a blackbody intrinsic flux.

Hershkowitz, Linder, and Wagoner ([46] and references therein) argue that the lack of a Balmer discontinuity in the spectra of SNe II can be used to set an upper limit to the electron density at the photosphere. The electron density is low enough that the electron-scattering opacity is predicted to be much greater than the absorptive opacity (such as provided by photoionization of hydrogen). Static plane-parallel model atmospheres for SNe II further predict that in the scattering-dominated case the photosphere forms at $\tau_{es} \gg 1$ and the intrinsic flux is "dilute", *i.e.*, less than the blackbody flux. From this Hershkowitz *et al.* conclude that when the blackbody approximation is made, the photometric angular size will be underestimated, the distance will be overestimated, and the result for H_o will be too small.

Taking the flux dilution into account will lead to a higher value of H_o *if* the dilution factor remains constant during the time interval of interest. It is worth stressing, however, that if the dilution factor varies with time, the effect on the distance can go in either direction. To illustrate this, suppose that the method is applied to observations made at just two times. In this case the distance can be expressed as

$$D = \frac{V_1 V_2 (t_2 - t_1)}{\frac{\theta_2 V_1}{\sqrt{\epsilon_2}} - \frac{\theta_1 V_2}{\sqrt{\epsilon_1}}} \tag{5}$$

where θ is the blackbody angular radius, ϵ is the flux dilution factor ($\epsilon \leq 1$), and the subscripts refer to the two times. If the dilution factor is constant, $D \propto \epsilon^{1/2}$. However, the color temperature decreases from $> 10,000K$ at maximum light to $\sim 6000K$ a month later. During these phases the ratio of electron scattering to absorptive opacity should be decreasing owing to recombination, so it is likely that $\epsilon_2 > \epsilon_1$. Table 2 illustrates that the factor by which the "blackbody" distance must be multiplied can be greater or less than unity, depending on the values of ϵ_1 and ϵ_2. Velocities ($V_1 = 9500$, $V_2 = 8000$ km s^{-1}) and photometric angular radii ($\theta_1 = 2.9 \times 10^{-11}$, $\theta_2 = 5.5 \times 10^{-11}$

radians) have been taken to be the values given by Branch *et al.* [41] for SN 1979c near maximum light (April 11) and at the time when the photosphere approached its maximum radius (May 28).

Table 2

Effect of Scattering on Distance to SN 1979c

	$\epsilon_2 =$ 1.0	0.8	0.6	0.4	0.2
$\epsilon_1 =$ 1.0	1.0†	-	-	-	-
0.8	1.1	0.9	-	-	-
0.6	1.3	1.0	0.8	-	-
0.4	1.7	1.3	0.9	0.6	-
0.2	11	3.4	1.7	0.9	0.4

†Ratio of true distance to "blackbody" distance.

In the atmospheres of SNe I, it is unlikely that the opacity is dominated by *electron* scattering. If the electron density is assumed to vary with radius as r^{-n}, the mass and kinetic energy *above the photosphere* can be expressed as

$$M(>R) = 0.027 \, \frac{n-1}{n-3} \, V_9^2 \, t_{15}^2 \, \mu_{es} \, \tau_{es}, \tag{6}$$

$$E_K(>R) = 0.027 \times 10^{51} \frac{n-1}{n-5} \, V_9^4 \, t_{15}^2 \, \mu_{es} \, \tau_{es} \tag{7}$$

where τ_{es} is the optical depth to electron scattering at the photospere and μ_{es} is the number of nucleons per free electron in the atmosphere. For SNe I at maximum light, characteristic values of $V = 12{,}000$ km s^{-1}, $t - t_o = 15$ days, $\mu_{es} = 10$ (doubly ionized carbon, oxygen, silicon, and sulphur), and $n = 7$ lead to $0.6\tau_{es}$ solar masses and $1.7 \times 10^{51} \tau_{es}$ ergs above the photosphere. An electron-scattering optical depth much in excess of unity would imply too much mass and kinetic energy. The conversion of 1.4 $M_\odot$ of carbon and oxygen to iron yields 2.2×10^{51} ergs; from this the binding energy of the white dwarf, about 0.5×10^{51} ergs, must be subtracted, so the maximum expected *total* kinetic energy is only 1.7×10^{51} ergs. Although the opacity in SNe I atmospheres apparently is not dominated by electron scattering, it *could* be dominated by scattering in bound-bound transitions, in which case the dilution arguments of Hershkowitz *et al.* might still apply.

In summary, the blackbody version of the Baade-Wesselink method, applied to both SNe II and SNe I, leads to low values of the Hubble constant, $40 \lesssim H_o \lesssim 60$ km s^{-1} Mpc^{-1}. It is not clear whether the result for H_o will increase or decrease when departures from the blackbody intrinsic flux are taken into account. Progress with this method requires the development of realistic model atmospheres for supernovae. The static plane-parallel atmospheres of Hershkowitz *et al.* [46] are a step in the right direction. Models of expanding, extended atmospheres such as those being developed by Harkness [47] will be needed if the Baade-Wesselink results for H_o are to become convincing.

IV. ^{56}Ni Radioactivity in SNe I

This approach is based upon a specific model of a Type I supernova. A carbon-oxygen white dwarf accretes matter from a companion star in a binary system. The accreted matter may be hydrogen or helium [48,49], subsequently converted to carbon and oxygen *via* shell burning near the white dwarf surface, or it may be carbon and oxygen in the first place, coming from a companion white dwarf [50-52]. In either case, as the mass of the white dwarf approaches the Chandrasekhar mass (1.4 $M_{\odot}$), nuclear burning of degenerate carbon occurs at the center of the star and generates a subsonic, convectively driven nuclear-burning front (a "deflagration") propagating outwards. The deflagration wave burns an inner portion of the white dwarf to nuclear statistical equilibrium (NSE), producing iron-peak elements, principally beta-unstable ^{56}Ni. Because the deflagration wave is subsonic, the outer parts of the star, ahead of the burning front, have time to begin to expand, and the decreasing density causes the deflagration to falter as it moves outwards. An intermediate portion of the star is burned to intermediate-mass elements, from oxygen to calcium, and an outer part of the star remains unburned carbon and oxygen. The nuclear energy overcomes the binding energy of the white dwarf and completely disrupts it; no neutron star is left behind. The initial radius of the white dwarf is so small that, in the absence of delayed heating of the gas by radioactivity, adiabatic cooling during expansion would lower the temperature before the supernova could achieve a large radiating surface, and the optical emission would be very dim. However, ^{56}Ni decays with a half life of 6.1 days to ^{56}Co, which decays in turn with a 77-day half life to stable ^{56}Fe. Partial trapping and thermalization of the energy carried by the decay products (gamma rays from ^{56}Ni, gammas and positrons from ^{56}Co) keeps the gas hot and powers the optical light curve [53-58].

The significance of this model for H_o is that the optical luminosity depends primarily on the amount of ^{56}Ni synthesized in the explosion. Arnett, Branch, and Wheeler [8] have outlined how the nickel mass and the value of H_o can be estimated. This approach is also discussed by Wheeler [59] in this volume. The following paragraphs merely outline the chain of reasoning leading from the carbon-deflagration model to H_o.

Arnett [60] has constructed analytical models for the light curves of SNe I. He finds that *at the time of maximum bolometric luminosity, the luminosity is equal to the instantaneous radioactive-decay luminosity, i.e.,* $L_{peak} = L_{peak}(M_{Ni}, t_1 - t_o)$, where M_{Ni} is the mass of ^{56}Ni synthesized by the explosion and $t_1 - t_o$ is the rise time to maximum (bolometric) light. This is an important prediction, because the rise time for SNe I can be estimated from observation, as discussed above. The dependence of the peak luminosity on the opacity, which is not well known, is absorbed in the observed rise time. To estimate the peak luminosity, then, we just need to know the nickel mass.

The nickel mass is determined by the velocity of the deflagration wave. The faster the deflagration, the more nickel produced. Unfortunately, the propagation of the deflagration front can not yet be calculated from first principles, so its velocity must be regarded as a free parameter. However, the nickel mass can be constrained by working backwards from its consequences. The nickel mass is related to the nuclear energy generated by the explosion, and the kinetic energy of the ejecta is to a good approximation given by the difference between the nuclear energy and the net binding energy. The kinetic energy in turn determines the velocity at the photosphere as a function of time, which can be inferred from analysis of SNe I spectra.

Synthetic spectra studies indicate that the velocity at the photosphere varies from 12,000 to 10,000 km s^{-1} during the first weeks after maximum light [25]. Hydrodynamic calculations can relate the velocities at the photosphere to the total kinetic energy. Sutherland and Wheeler [57] find that 1.1×10^{51} ergs is required. The net binding energy of the white dwarf is 0.5×10^{51} ergs, so the nuclear energy is 1.6×10^{51} ergs. If the nuclear burning produced pure ^{56}Ni, this would require $M_{Ni} \simeq 1.0$ $M_\odot$. However, the production of isotopes other than ^{56}Ni in the NSE region, and of intermediate-mass elements (O-Ca) farther out, must be taken into account. Models computed by Nomoto, Thielemann, and Yokoi [61] and Woosley, Axelrod, and Weaver [58], with detailed treatments of nucleosynthesis in the deflagrations, show that $M_{Ni} \simeq 0.6$ $M_\odot$ is required.

Adopting $M_{Ni} = 0.6$ $M_\odot$ and a rise time to *bolometric* maximum of 17 days (corresponding to a rise time to *blue* maximum of 15 days [57]) gives $L_{peak} = 1.4 \times 10^{43}$ ergs s^{-1}. The conversion from a bolometric luminosity to an absolute blue magnitude appears to be straightforward for SNe I. The energy distribution at peak light resembles a $20,000K$ blackbody truncated near λ4000Å. When the theoretical bolometric luminosity is distributed over the spectrum in this way, the value of M_B becomes -19.5. Combining this with *eqn.* (2) (and allowing for a difference of 0.1 *mag.* between 15 and 17 days) yields $H_o = 58$ km s^{-1} Mpc^{-1}.

The dominant uncertainty is in the value of the nickel mass. The dependence of H_o on M_{Ni} is

$$H_o = 45\left(\frac{M_{Ni}}{M_\odot}\right)^{-1/2}. \tag{8}$$

In the white dwarf model, a rigorous upper limit is $M_{Ni} \leq 1.4$ $M_\odot$, corresponding to $H_o \geq 38$. But this is an extreme limit, since optical spectroscopy establishes the presence of intermediate-mass elements in the ejecta (as does X-ray spectroscopy of Tycho's remnant [62,63] if Tycho's SN really was a Type I). A more realistic upper limit is $M_{Ni} \lesssim 1.0$ $M_\odot$, and $H_o \gtrsim 45$.

The *minimum* kinetic energy, considering the uncertainties in establishing the velocities from spectroscopy, is 0.6×10^{51} ergs [57], corresponding to $M_{Ni} \gtrsim 0.4$ $M_\odot$ and $H_o \lesssim 71$; this may be a conservative lower limit, because it probably corresponds to too much unburned carbon and oxygen. $H_o \approx 100$ would correspond to $M_{Ni} \approx 0.2$ $M_\odot$. Such a weak explosion would barely disrupt the white dwarf, and the velocities at the photosphere would be much too low.

If the model of SNe I is correct, this is a relatively clean estimate of H_o. It is important to emphasize that in this approach, one does *not* use an observed color temperature to predict an absolute emissivity, as one must do to apply the Baade-Wesselink technique. It will be important to apply the ^{56}Ni method to individual SNe I beyond the local supercluster, because H_o could be overestimated here due to differences between the relatively nearby SNe I, whose characteristics define the model parameters, and the remote SNe I used to establish *eqns.* (1) and (2). The lower limit to H_o, based on the Chandrasekhar mass, should already be secure.

V. Summary and Conclusion

In their simplest forms, all three of the approaches to H_o outlined here suggest that the value of the Hubble constant is relatively low, nearer to 50 than to 100 km s^{-1} Mpc^{-1}. If the model on which it is based is correct, the ^{56}Ni radioactivity method appears to be the most powerful. Adopting the less conservative limits to the nickel mass given above, and now considering the other smaller sources of error, the ^{56}Ni method gives $40 \lesssim H_o \lesssim 70$ km s^{-1} Mpc^{-1}.

It may be useful to turn the argument around and ask what the implications for supernovae would be, if the Hubble constant should be demonstrated to be more like 100 km s^{-1} Mpc^{-1}, as implied by recent determinations based on the bulk properties of galaxies [64,65]. For the standard candle approach, it would mean that NGC 5253 and IC 4182, which have produced the brightest normal SNe I that we have seen, are both at distances of ~ 2 Mpc, coincidentally projected in front of more distant groups. The blackbody Baade-Wesselink method would be in error for both SNe I and SNe II, presumably due to dilution effects in scattering-dominated atmospheres; the sources of the scattering in SNe I and SNe II would have to be different. The implication for the ^{56}Ni radioactivity picture would be that SNe I must leave bound remnants, after all. If the remnant is a white dwarf, the ejected mass is less than 1.4 $M_{\odot}$; the kinetic energy needed to account for the velocities at the photosphere, and therefore the nickel mass, is reduced. If a neutron star is formed (contrary to current observational evidence [66-68]), kinetic energy and nickel mass become uncoupled because some kinetic energy can come from the binding energy of the neutron star. Perhaps none of these accomodations to $H_o \approx 100$ are completely out of the question, but with supernovae (as with cosmology) the picture is simpler if $H_o \approx 50 - 60$ km s^{-1} Mpc^{-1}.

I am grateful for helpful discussions with David Arnett, Ken'ichi Nomoto, Craig Wheeler, and Stan Woosley, and for the support and hospitality of the Lick Observatory during the fall of 1984. This work has been supported by NSF grant AST 82-18625.

REFERENCES

[1] Tammann, G. A. 1982, in *Supernovae: A Survey of Current Research,* ed. M. J. Rees and R. J. Stoneham (Dordrecht, Reidel Publishing Company), p. 371.

[2] Falk, S. W., and Arnett, W. D. 1977, *Ap. J. Suppl.,* **33**, 515.

[3] Chevalier, R. A. 1976, *Ap. J.,* **207**, 872.

[4] Weaver, T. A., and Woosley, S. E. 1980, *Ann. N. Y. Acad. Sci.,* **336**, 335.

[5] Litvinova, I. Y., and Nadyozhin, D. K. 1983, *Ap. Sp. Sci.,* **89**, 89.

[6] Trimble, V. 1984, *J. Ap. Astr.,* in press.

[7] Sandage, A. R. 1985, this volume.

[8] Arnett, W. D., Branch, D., and Wheeler J. C. 1984, submitted to *Nature.*

[9] Rust, B. W. 1974, Ph.D Thesis, University of Illinois.

[10] Pskovskii, Y. P. 1977, *Soviet Astr. - AJ,* **21**, 675.

[11] Branch, D. 1982, *Ap. J.,* **258**, 35.

[12] Panagia, N., *et al.* 1984, preprint.

[13] de Vaucouleurs, G. 1984, preprint.

[14] Baade, W. 1943, *Ap. J.,* **97**, 119.

[15] Baade, W. 1945, *Ap. J.,* **102**, 309.

[16] van den Bergh, S. 1970, *Nature,* **225**, 503.

[17] Clarke, D. H., and Stephenson, F. R. 1977, *The Historical Supernovae* (Oxford: Pergamon Press).

[18] Pskovskii, Y. P. 1978, *Soviet Astr. - AJ,* **22**, 420.

[19] Barbon, R., Ciatti, F., and Rosino L. 1979, *Astr. Ap.,* **72**, 287.

[20] Doggett, J. B. 1984, submitted to *A. J.*

[21] de Vaucouleurs, G. 1984, preprint.

[22] Hoffleit, D. 1939, *Harvard Bull.,* No. **910**, p. 6.

[23] Johnson, W. A. 1936, *Harvard Bull.,* No. **902**, p. 11.

[24] Kirshner, R. P., Oke, J. B., Penston, M. V., and Searle, L. 1973, *Ap. J.,* **185**, 303.

[25] Branch, D., Lacy, C. H., McCall, M. L., Sutherland, P. G., Wheeler, J. C., and Wills, B. J. 1983, *Ap. J.,* **270**, 123.

[26] Baade, W., and Zwicky, F. 1938, *Ap. J.,* **88**, 411.

[27] Minkowski, R. 1939, *Ap. J.,* **89**, 156.

[28] Wild, P. 1960, *P. A. S. P.,* **72**, 97.

[29] Branch, D. 1972, *Astr. Ap.,* **16**, 247.

[30] Maza, J., and van den Bergh, S. 1976, *Ap. J.,* **204**, 519.

[31] de Vaucouleurs, G., de Vaucouleurs, A., and Corwin, H. G. 1976, *Second Reference Catalogue of Bright Galaxies* (Austin: University of Texas Press).

[32] Sandage, A. R., and Brucato, R. 1979, *A. J.,* **84**, 472.

[33] de Vaucouleurs, G. 1975, in *Galaxies and the Universe, Vol. IX of Stars and Stellar Systems,* ed. A. Sandage, M. Sandage, and J. Kristian (Chicago: University of Chicago Press), Ch. 14, p. 557.

[34] de Vaucouleurs, G. 1979, *A. J.,* **84**, 1270.

[35] Sandage, A. R., and Tammann, G. A. 1974, *Ap. J.,* **194**, 559.

[36] van den Bergh, S. 1980, *P.A.S.P.,* **92**, 122.

[37] Sandage, A. R., and Tammann, G. A. 1982, *Ap. J.,* **256**, 339.

[38] de Vaucouleurs, G. 1979, *Ap. J.,* **227**, 729.

[39] Baade, W. 1926, *Astr. Nachr.,* **228**, 359.

[40] Wesselink, A. 1985, this volume.

[41] Branch, D., Falk, S. W., McCall, M., Rybski, P., Uomoto, A. K., and Wills, B. J. 1981, *Ap. J.*, **244**, 780.

[42] Kirshner, R. P., and Kwan, J. 1974, *Ap. J.*, **193**, 27.

[43] Arnett, W. D. 1982, *Ap. J.*, **254**, 1.

[44] Barbon, R., Ciatti, F., and Rosino L. 1973, *Astr. Ap.*, **25**, 241.

[45] Branch, D., and Patchett B. 1973, *M.N.R.A.S.*, **161**, 71.

[46] Hershkowitz, S., Linder, E., and Wagoner, R. V. 1984, preprint.

[47] Harkness, R. 1985, this volume.

[48] Nomoto, K. 1980, in *Type I Supernovae*, ed. J. C. Wheeler (Austin: University of Texas Press), p. 164.

[49] Weaver, T. A., Axelrod, T. S., and Woosley, S. E. 1980, in *Type I Supernovae*, ed. J. C. Wheeler (Austin: University of Texas Press), p. 113.

[50] Iben, I. Jr., and Tutukov, A. V. 1984, *Ap. J. Suppl.*, **55**, 335.

[51] Webbink, R. F. 1984, *Ap. J.*, **277**, 355.

[52] Paczynski, B. 1983, preprint.

[53] Colgate, S. A., and McKee, C. 1969, *Ap. J.*, **157**, 623.

[54] Arnett, W. D. 1979, *Ap. J. (Letters)*, **230**, L37.

[55] Chevalier, R. A. 1981, *Ap. J.*, **246**, 267.

[56] Schurmann, S. R. 1983, *Ap. J.*, **267**, 779.

[57] Sutherland, P. G., and Wheeler, J. C. 1984, *Ap. J.*, **280**, 282.

[58] Woosley, S. E., Axelrod, T. S., and Weaver, T. A. 1984, in *Stellar Nucleosynthesis*, ed. C. Chiosi and A. Renzini (Dordrecht: Reidel Publishing Company), p. 263.

[59] Wheeler, J. C. 1985, this volume.

[60] Arnett, W. D. 1982, *Ap. J.*, **253**, 785.

[61] Nomoto, K., Thielemann, F.-K., and Yokoi, K. 1984, *Ap. J.*, **286**, in press.

[62] Shull, J. M. 1982, *Ap. J.*, **262**, 308.

[63] Sarazin, C., private communication.

[64] Aaronson, M., Mould, J., Huchra, J., Sullivan, W. T., III, Schommer, R. A., and Bothun, G. D. 1980, *Ap. J.,* **239,** 12.

[65] de Vaucouleurs, G. 1983, *Ap. J.,* **268,** 468.

[66] Helfand, D. J. 1980, in *Type I Supernovae,* ed. J. C. Wheeler (Austin: University of Texas Press), p. 20.

[67] Nomoto, K., and Tsuruta, S. 1981, *Ap. J. (Letters)* **250,** L19.

[68] Helfand, D. J., and Becker, R. H. 1984, *Nature,* **307,** 215.

TYPE I SUPERNOVAE AS STANDARD CANDLES

R. Cadonau[1], A. Sandage[2], and G. A. Tammann[1,3]

1 Astronomisches Institut der Universität Basel

2 Mount Wilson and Las Campanas Observatories

3 Visiting Associate Mount Wilson and Las Campanas Observatories; Associate European Southern Observatory

Abstract

Observations are compiled to construct a mean light curve of SNe I in different colors. Individual SNe I show generally no systematic deviations from these templet light curves, and occasional deviations are explained as photometric errors, which can be quite severe in the case of SNe. The peak luminosity of absorption-free SNe I is also uniform with an intrinsic rms scatter of <0.3 mag. The calibration of the peak luminosity yields $M_B(\max) = -20.0 \pm 0.4$, which requires $H_0 = 43 \pm 10$ km s^{-1} Mpc^{-1}. SNe I are probably among the best standard candles known and hence have in view of their high luminosity important applications for distance determinations and for cosmological tests to be performed with Space Telescope.

I. Introduction

It is generally agreed that supernovae of type II (SNe II) have light curves which differ in shape and in absolute maximum magnitude. The situation is not so clear for supernovae of type I (SNe I). While some authors have suggested that SNe I have a significant range in their decline rates, e.g. as measured by the time needed for a SNe I to decline by 2 mag. after maximum, others have argued that SNe I light curves are templets which repeate from one object to the other to within the observational errors. A dichotomy of light curves according to their behavior during the first ~ 30 days after maximum was suggested, for instance, by Barbon et al. (1973; Barbon 1980) who devided them into "fast" and "slow" SNe I. A more continuous range of decline rates was advocated by Rust (1974), Pskovskii (1977) and Branch (1982). These authors found also correlations, though weak, between decline rate and maximum luminosity. On the other hand Kowal (1968), Tammann (1978), and Elias et al. (1981) have stressed the great similarity of well observed light curves, which would leave little room for intrinsic variations. In view of this unsatisfactory situation we re-investigate here the question whether SNe I are standard candles, including also more recent observations.

After a presentation of the composite light curve of SNe I in Section II, we discuss individual deviations from this mean curve in Section III. In Section IV we determine an upper limit for the intrinsic scatter of the absolute magnitude at maximum light and in Section V we attempt a calibration of the absolute maximum magnitude M_B(max). Some conclusions are compiled in Section VI.

II. The Composite Light Curve of SNe I

A literature search has resulted in a nearly complete set of optical magnitude determinations of SNe I. The quality of the data is quite variable. While the older values are given mainly in the pg system, often without specifications as to the emulsion and filters used and as to the origin of the magnitudes of the comparison stars, the newer photographic or photoelectric observations are generally in the UBV system. The main systematic errors of the available photometry stem from the following three sources: (1) zero point errors of photographic pg and UBV magnitudes, which can be particularly important in cases where the local standard magnitudes were obtained from magnitude transfers; (2) scale errors of the local standard magnitudes used, which translate necessarily into a distortion of the <u>form</u> of the SN light curve; and (3) the interaction of the SN with the background light of the parent galaxy, which becomes more severe as the SN becomes fainter and hence again affects the <u>form</u> of the light curve and moreover introduces important variations between observers using different instrumentation. The degree of background contamination depends on the radial distance of the SN from the center of the parent galaxy and on the technique employed (particularly severe for iris photometry and photoelectric measurements). These error sources together with the facts that the photometry extends typically down to the technically feasable brightness level and that the photometry problem was quite occasionally treated in some of the original sources, lead <u>a priori</u> to the expectation that the scatter of SN magnitudes due to observational errors is considerably larger than the one which observers would normally accept for more conventional objects. With this warning in mind we proceed with the construction of the composite light curve, treating pg and B magnitudes separately, for reasons which will become evident later.

For 16 and 22 SNe I the available pg and B magnitudes, respectively, are sufficiently numerous and cover a sufficient time interval that they can contribute to the definition of a mean lightcurve. The magnitudes of each SN were plotted against time and the resulting lightcurves were shifted horizontally (corresponding to an arbitrary choice of luminosity and/or distance) such as to obtain a mean curve with minimum scatter. The result for the pg and B data is shown in Fig. 1. The pg curve is replotted on a compressed time scale in Fig. 2 in order to include observations up to 500 days after maximum.

The quite well defined curves in Fig. 1 and 2 confirm the earlier discovery of Barbon et al. (1973), viz. that SNe I follow light curves which are surprisingly similar in shape. But while the scatter about the mean light curve allowed these authors to distinguish between two subtypes, "fast" and "slow", it can be seen in Fig. 1 that the scatter is mainly caused by the pg data, which taken separately have a mean epoch of 1957 and a mean deviation from the mean line of $\sigma(m) \approx 0.35$ mag. The corresponding value for the B light curve with a mean epoch 1970 is only $\sigma(m) \approx 0.18$ mag. near maximum and $\sigma(m) \approx 0.29$ mag. fifty days later, and because the B magnitudes are on the average the more modern data, their smaller scatter is probably a more realistic upper limit to the true scatter. This narrows the basis on which the existence of subtypes and different decline rates of SNe I could be postulated. Also the mean light curves of U and V magnitudes (which are not shown here) agree with this conclusion, although fewer data points are available in these colors. The near identity of the light curves was furthermore demonstrated by infrared photometry of three SNe I which agree for $\gtrsim 100$ days to within $\lesssim 0.1$ mag. (Elias et al. 1981).

The epoch of maximum is clearly wavelength-dependent. It was therefore tested whether the epoch of B maximum t (B_{max}) and pg maximum t (pg_{max}) coincide. Only three SNe I are available for this test. They show indeed that the times of maxima agree to within $\pm$ 0.5 days. The shape of the mean B and pg light curves agree so well that it was assumed in Fig. 1 that they are identical. Ten SNe I are available to determine the weighted magnitude difference (B-pg) to be 0.28 ± 0.05 mag.

From a detailed discussion of the rising branch of the light curve, to be published elsewhere, we conclude that explosion must occur at t $(B_{max}) < -20$ days. The light curve remains for 25 days within 1 mag. from maximum. Between $7 < t\ (B_{max}) < 30$ days the mean decline rate is 0.085 mag./day to turn then gradually into a more or less exponential decline; after t $(B_{max}) = 42$ up to $\sim$ 500 days the latter is approximated by 0.017 mag./day. From inspection of Fig. 1 and 2 this value may be deemed too large because the adopted "mean" light curve at later phases passes below the center of weight of the observations. The present fit was adopted because much weight was given to Baade's still unmatched SN 1937c in IC 4182 and because the observational errors tend to produce too bright values at faint levels due to

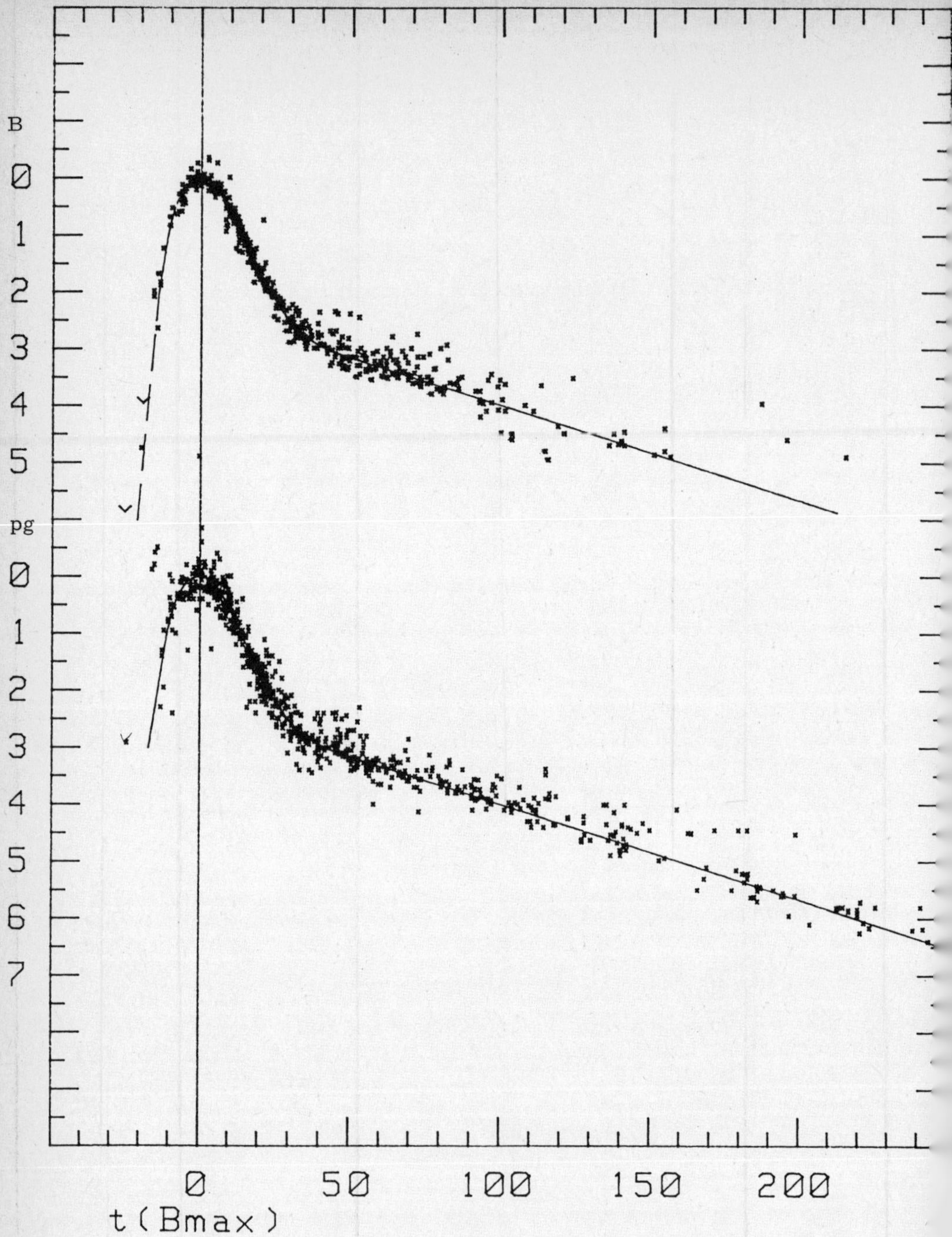

Fig. 1. The composite pg light curve for 16 SNe I (lower curve) and the composite B light curve for 22 SNe I. The adopted templet light curves are shown as full lines. Note the smaller scatter in B than in pg.

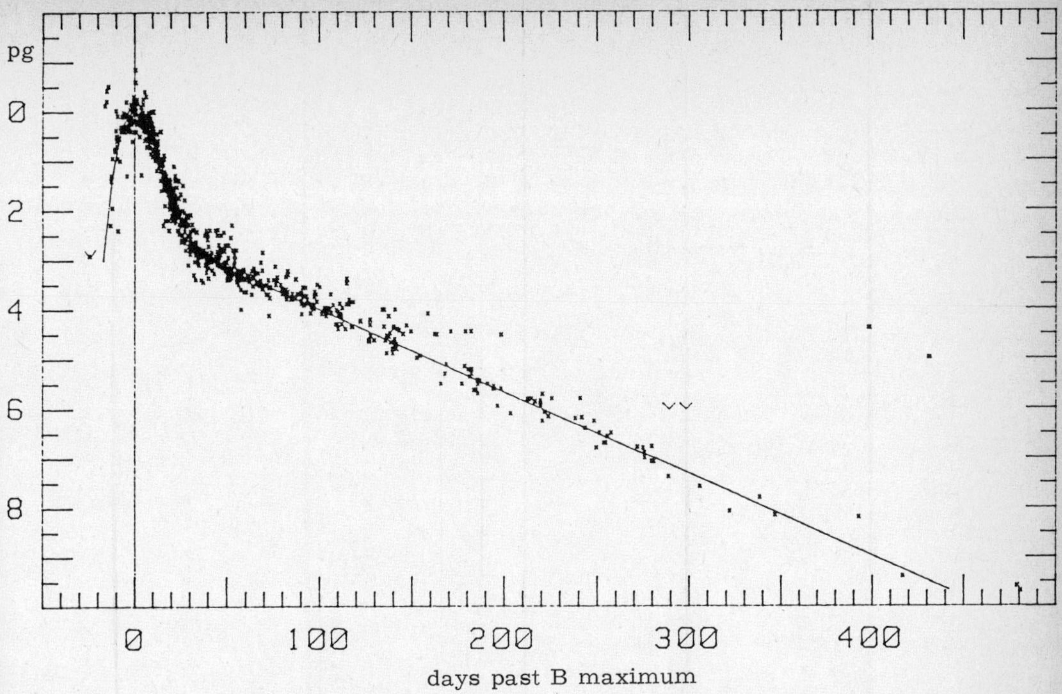

Fig. 2. The composite pg light curve of 16 SNe I. Same as Fig. 1, lower curve, but here extending to almost 500 days.

background light, which may become particularly severe for photoelectric observations. For this reason the decline rate during the exponential phase should be taken only as an approximation.

SNe I can be distinguished from SNe II unambiguously only from spectroscopy. Not for all objects in Fig. 1 and 2 spectroscopy is available. But most objects lacking spectroscopy have occurred in E galaxies and this is generally taken to be sufficient evidence for a type I assignment, because all spectroscopically investigated SNe in E galaxies have so far proven to be of type I. It is, however, possible that Fig. 1 and 2 still contain a few objects which are not genuine SNe I. This possibility can only strengthen our conclusion, that true SNe I define a templet light curve which they follow to within the observational error.

III. A Comparison of the Templet Light Curve with Individual SNe I

The photometric data for over 60 SNe I were compared with the pg or/and B templet light curves from Fig. 1 and 2 and - where warranted by the available data - also with the V and U templets. No deviations were found which could not be explained as the result of observational errors. A few illustrating examples are given in the following.

The pg, B, V, and U observations of two SNe I, which were assigned to the "fast" subtype by Barbon et al. (1973), are shown in Fig. 3 together with the corresponding templet light curves as adopted here. It should be noted that these templets leave in their relative horizontal position no freedom because the time differences between the different maxima are quite well determined, i.e. $t(B_{max}) - t(V_{max}) = 2$ days and $t(U_{max}) - t(B_{max}) = -3$ days. The left panel of Fig. 3 illustrates the case of SN 1971 i. The pg observations of Deming et al. (1973) are indeed quite bright shortly after maximum and could be taken to indicate a "fast" SN I. However, the presumably better B magnitudes of the same authors fall almost exactly on the B templet. In B light there are additional data of Barbon et al. (1973a); these show a systematic scale difference from those of Deming et al. and are quite bright near maximum. To conclude from this that SN 1971 i was "fast" would imply, however, to discard the B magnitudes of Deming et al. In V light the magnitudes of Deming et al. and Barbon et al. fall sufficiently close to the V templet that the conclusion of a "fast" SN I seems unwarranted.

The right panel of Fig. 3 shows the case of SN 1967 c, which according to Barbon et al. (1973) again is a "fast" SN I. The magnitudes of various authors fall on the pg, B, V (or pv) and U templets as well as one can expect. The object has hence all photometric properties of a normal SN I.

In Fig. 4 two typical examples are shown of what Barbon et al. (1973) have classified as "slow" SNe I, i.e. SN 1957 b and SN 1969 c. The reader may judge for himself whether the observations give any evidence for a slower decline than the templets require.

A detailed check of the light curves of all SNe I with photometric data led again and again to the same result: significant deviations from the templets are either non-existing or at least quite rare.

An apparently alarming exception is the case of SN 1939 b which, having occurred in an E galaxy, belongs most likely to type I. It declined by ~ 3.5 mag. within only 30 days after maximum, while the templet requires 75 days for such a decline.

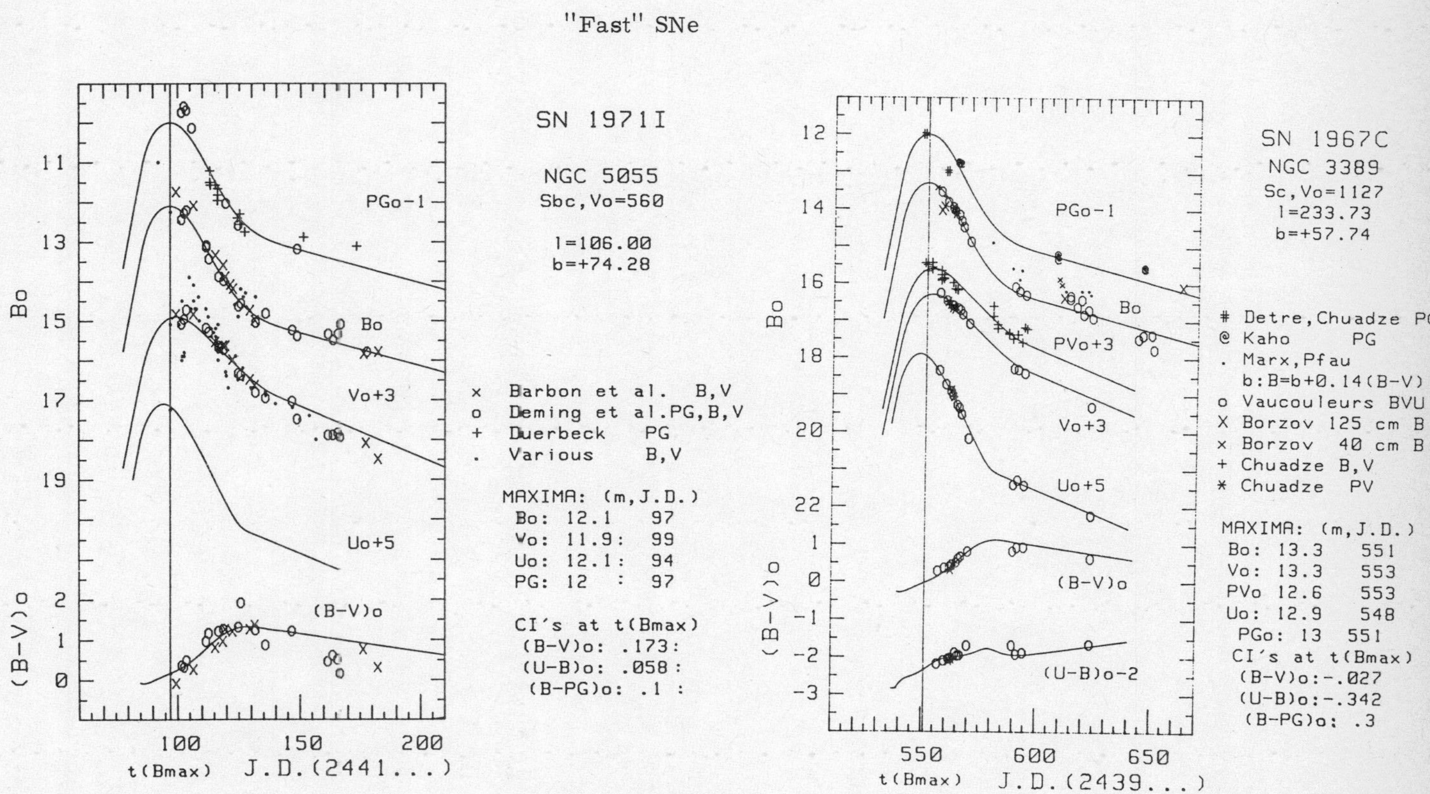

Fig. 3. Two so-called "fast' SNeI. The templet light curves are fitted to the available observations. The key to the abbreviated sources can be found in Flin et al. (1979).

"Slow" SNe

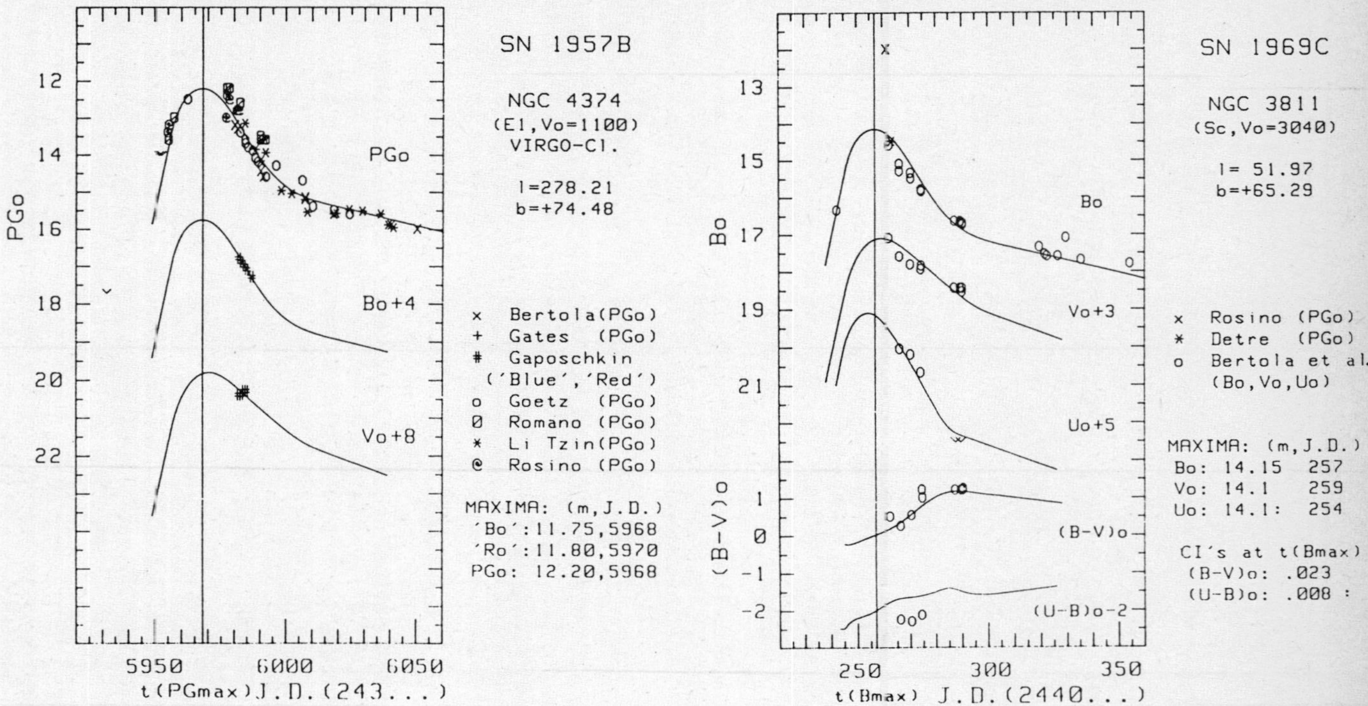

Fig. 4. Two so-called "slow" SNe I. The templet light curves are fitted to the available observations. The key to the abbreviated sources can be found in Flin et al. (1979).

However, Baade's decisive photometry has never been published by himself, and his data were presented only in a smoothed small-scall light curve (Minkowski 1964). The case of this SN carries therefore relatively little weight.

The fastest light curve of all SNe was found for SN 1885 (S And) in M 31 (Jones 1976; de Vaucouleurs and Buta 1981). The weak spectroscopic evidence for this object, as compiled by Jones (1976), gives, however, no assurance that the object was of type I.

In summary of this Section we note that the great majority of all SN light curves tested follow the templets impressively well, and there is not a single bona fide type I SN whose deviations from the templets could not be explained by plausible observational errors.

IV. Are SNe I Standard Candles at Maximum?

The similarity of the shape of the light curves of SNe I, as discussed in the previous Section, suggests that they may reach at maximum a uniquely defined absolute magnitude.

Unfortunately this expectation can be checked only with SNe I which occurred in E/S0 galaxies and which presumably suffer minimum absorption in their parent galaxies. SNe I in spiral and irregular galaxies are systematically fainter and redder, which must be taken as a clear indication for internal absorption (Tammann 1982). The 12 SNe I in E galaxies with the best determined maxima are listed in Table 1. Unfortunately E galaxies are very slow producers of SNe and no such SN was photometrically observed after 1972; modern high-quality photometry of some SNe I in E galaxies would be most desirable.

Table 1 lists the respective SNe I in chronological order. Column 2 gives the parent galaxies and their morphological type. Their recession velocities, corrected for the appropriate Virgocentric infall vector (Kraan-Korteweg 1985), are shown in column 3. The absolute magnitudes $M_B^o(\max)$ in column 5 are calculated from the apparent maximum magnitudes in column 4 on the assumption of H_o = 50 km $s^{-1}Mpc^{-1}$. Column 6 indicates those cases where the B maximum has been inferred from the pg system and/or where the parent galaxy is a cluster member.

The scatter of the $M_B^o(\max)$ values in Table 1, column 5, amounts to $\sigma(M) = 0.45$ mag. Much of this scatter is certainly due to observational errors. Considering that some of the light curves are defined by only few observations and that some maximum magnitudes require considerable extrapolation, we estimate the observational errors - probably conservatively - to be 0.3 mag. The depth effect of the

Table 1

SNe I in E/S0 galaxies with reasonably determined maxima

SN	Galaxy		v_{corr}	$B^o(max)$	$M_B^o(max)$	Remarks
(1)	(2)		(3)	(4)	(5)	(6)
1919a	N 4486	E	1187	11.9	-19.8	pg, Virgo
1939a	N 4636	E/S0	1187	12.7	-19.0:	pg, Virgo?
1957b	N 4374	E1	1187	12.48	-19.22	pg, Virgo
1960r	N 4382	S0pec	1187	11.89	-19.81	Virgo
1961d	Anon	E/S0	7217	16.53	-19.27	pg, Coma
1961h	N 4564	E6	1187	11.48	-20.22	pg, Virgo
1962a	Anon	E/S0	7217	15.63	-19.55	pg, Coma
1962a	Anon	E1	14455	17.4	-19.9	pg
1965i	N 4753	S0pec	1187	12.56	-19.14	Virgo
1968a	N 1275	Epec	5342	14.73	-20.41	pg
1970j	N 7619	E3	3881	14.35	-20.10	Pegasus
1972j	N 7634	E	3374	14.35	-19.80	
					-19.69 ± 0.13	
					$\sigma(M) = 0.45$	

Virgo cluster members introduces a scatter of ~ 0.2 mag. Some SNe I may suffer still from internal absorption, as in the case of SN 1965i, which occurred in NGC 4753, an S0 pec galaxy. This possibility becomes particularly evident in the case of the very red and faint SN 1983g, which occurred in the same galaxy and which matches perfectly the templet light curves. Compounding these external error sources leaves an intrinsic scatter of $\sigma(M) \lesssim 0.3$ mag. But if we have underestimated the observational errors the true intrinsic scatter may be vanishingly small.

In fact, an intrinsic scatter of ~ 0.5 mag. would introduce a systematic bias in the sense that SNe I with higher recession velocities would be intrinsically brighter. The absence of such a Malmquist bias gives independent evidence for the scatter being small (cf. Tammann 1982).

The strongest evidence, however, comes from JHK photometry of the two SNe I which occurred in NGC 1316 (Elias et al. 1981). The infrared magnitudes minimize the effect of internal absorption in this peculiar galaxy and they demonstrate that the two SNe I had identical luminosities to within 0.1 mag.!

The repetitiveness of SNe I luminosities is only plausible if also their spectra follow a standard pattern. Several optical and UV spectroscopists have indeed commented on the striking similarity of their spectral properties (Oke and Searle 1974; Branch 1984; Panagia 1982).

Finally, also statistical and theoretical considerations are compatible with SNe I being standard candles. It is obvious that only a minute fraction of all stars in a galaxy become ever SNe I, and these are presumably stars with highly specific characteristics. Current models involving white dwarfs, which accrete matter in an evolving binary system until their mass approaches the Chandrasekhar limit, would explain and indeed require the rarity and uniformity of SNe I.

V. The Absolute Maximum Magnitude of SNe I

The result of the previous Section, that SNe I are (nearly) perfect standard candles, calls for the absolute magnitude calibration of this standard candle.

For two parent galaxies of SNe I distances are available from photometry of their brightest stars (Sandage and Tammann 1982). The relevant data are compiled in Table II.

Table II

Distances of two nearby galaxies with SNe I

SN (1)	Galaxy (2)	$(m-M)^o$ (3)	$m_B(max)$ (4)	E_{B-V} (5)	A_B (6)	$M_B(max)$ (7)
1937 c	IC 4182	28.21	8.78	0.32	0.64	-20.07
1954 a	N 4214	28.92	9.78	0.50	1.00	-20.14

The apparent maximum magnitudes in column 4 are newly determined from fits of the available photometry to our templet light curves. The color excesses E_{B-V} of the SNe I in column 5 follow from the observed colors (B-V) and an adopted intrinsic color of $(B-V)^o = -0.27$ at $t\ (B_{max}) = 0$. The latter value is determined from three SNe I which occurred in E galaxies and for which photometry in B (or pg) and V is available. The blue absorption A_B in column 6 is calculated from E_{B-V} with an assumed absorption-color excess ratio $R = A_B/E_{B-V} = 2$. This unconventionally low value seems appropriate on the average for all outlying SNe I in spiral and irregular galaxies. The value of $R \approx 4$, which holds for the solar neighborhood, would make reddened SNe I overluminous and would yield a correspondingly low value of H_o. A variation of R with radial distance from the galaxian center has been found also on independent grounds (Searle 1983; Iye and Richter 1984).

The conclusion from Table II is that SNe I reach an absolute blue maximum magnitude of $M_B(\max) = -20.1 \pm \sim 0.4$. This directly calibrated value finds support via four different routes:

1) The only two historical SNe, which can be assigned to type I with reasonable confidence, are SN 1572 (Tycho's) and SN 1607 (Kepler's). A compilation of their observational parameters results in absolute visual maximum magnitudes of $M_V(\max) = -19.51$ and -19.98, respectively (Tammann 1982). Unfortunately the distance of 4 kpc used for both objects is still uncertain. This is illustrated by the Tycho SN distance of 2.3 kpc and the Kepler SN distance of 4.4 kpc, recently proposed by Green (1984), which would require $M_V(\max) = -18.31$ and -20.19. However, the 1.8 mag. difference between these two values is not compatible with the present result that SNe I are standard candles. Taking the distance of 2.3 kpc for Tycho's SN at face value (cf. de Vaucouleurs 1984) would therefore require an even smaller distance for Kepler's SN, which not only would seem very improbable, but also would result in an alarmingly high Galactic surface density of historical SNe. Moreover the small distance of Tycho's SN conflicts with the <u>lower</u> distance limit of 4 kpc obtained by Schwarz et al. (1980), which implies for this object $M_V(\max) < -19.51$. The best compromise for the historical SNe I would therefore seem to be $< M_V(\max) > = -19.75$ and - with $(B-V)^o = -0.27$ - $M_B(\max) = -20.0$, but because of the poorly known distances the result carries still an uncertainty of $\sim \pm 0.6$ mag.

2) Purely physical, yet model-dependent expansion parallaxes of SNe I yield $M_B(\max) = -19.8 \pm 0.7$ (Arnett 1982; Branch 1982).

3) An expansion parallax, following the Baade-Wesselink method, was determined for a type II (!) SN in the Virgo cluster galaxy NGC 4321 (Branch et al. 1981). The resulting cluster distance of 23 ± 3 Mpc, combined with the mean apparent magnitude $< m_B(\max) > = 12.06$ of the five certain Virgo cluster SNe I of Table I, yields $M_B(\max) = -19.75 \pm 0.30$. Also this route via SNe II is, of course, dependent on the adopted model atmosphere, but it is reassuring to note that the expansion parallaxes of SNe I and SNe II yield fully consistent results.

4) Current SNe I models of exploding carbon-oxygen white dwarfs invoke the radioactive decay of Ni^{56} to explain the light curves. The amount of Ni^{56} to accelerate enough material to the observed expansion velocities and to provide an estimated total kinetic energy of 10^{51} ergs is certainly more than $0.4\,\mathfrak{M}_\odot$ and probably $\gtrsim 0.7\,\mathfrak{M}_\odot$; this amount of fuel provides a peak luminosity of at least $M_B(\max) = -19.0$, or

rather -19.6 (Branch 1982; Sutherland and Wheeler 1983).

A value of $M_B(\max) = -20.0 \pm 0.4$ for SNe I is therefore in good agreement with all available evidence. In Section IV the absolute magnitude of SNe I was found to be

$$M_B(\max) = -19.69 \pm 0.13 + 5 \log (H_o/50). \qquad (1)$$

Inserting the preferred value of $M_B(\max) = -20.0 \pm 0.4$ into equation (1) and solving for H_o yields

$$H_o = 43^{+10}_{-7} \text{ km s}^{-1}\text{Mpc}^{-1}.$$

VI. Conclusions

- There is no photometric evidence for light curve variations of SNe I. Deviations of individual SNe I from the templet light curves are generally smaller for more modern observations and can be fully explained by plausible assumptions on the observational errors. In particular there is no basis for "fast" and "slow" SNe I which would differ in their decline rates during the first 30 days after maximum.
- SNe I in E galaxies, i.e. those which suffer minimal absorption in their parent galaxies, have constant absolute maximum magnitudes within rms errors of $\lesssim 0.3$ mag. Judging from infrared photometry the true intrinsic luminosity variations at maximum (or at any other fixed phase) are still much smaller. SNe I are therefore nearly perfect standard candles.
- The value of $M_B(\max)$ can be calibrated with two SNe I in nearby field galaxies whose distances are known from the photometry of their brightest stars; the result finds further support from historical SNe I, expansion parallaxes of SNe I and SNe II, and from the necessary Ni^{56} production in explosive carbon-oxygen white dwarf models. $M_B(\max) = -20.0 \pm 0.4$ is adopted as the best value.
- The adopted value of $M_B(\max)$ requires a Hubble constant of 43^{+10}_{-7} km s^{-1}Mpc^{-1}. This value is freed from the effects of Virgocentric streaming velocities and reflects therefore the global expansion rate. A rounded value of $H_o = 50$ is discussed in the light of independent evidence elsewhere in this volume.
- SNe I used as standard candles have important advantages over brightest cluster galaxies, i.e. their luminosity evolution is expected to be small, they do not suffer from dynamical evolution, and their photometry is easier than for extended objects. They can play an important role in cosmology if they are observed with Space Telescope out to $z \approx 0.5$. The resulting Hubble diagram would provide a unique chance to confine the possible range of the deceleration parameter q_o and of the

cosmological constant Λ (cf. Sandage and Tammann 1984), and the expected time dilation of their light curves by a factor (1+z) would provide a fundamental test for the Doppler nature of the redshifts observed in distant galaxies (cf. Tammann 1979).

Partial support from the Swiss National Science Foundation is gratefully acknowledged. We thank Mrs. M. Saladin for the typing of the manuscript.

References

Arnett, W.D. 1982, in: Supernovae: A Survey of Current Research, eds. M.J. Rees and R.J. Stoneham, Dordrecht: Reidel, p. 221.

Barbon, R. 1980, in: Type I Supernovae, ed. J.C. Wheeler, Austin: University of Texas, p. 16.

Barbon, R., Ciatti, F., and Rosino, L. 1973, Astron. Astrophys. 25, 241.

Barbon, R., Ciatti, F., and Rosino, L. 1973a, Mem. Soc. Astron. Italiana 44, 65.

Branch, D. 1982, in: Supernovae: A Survey of Current Research, eds. M.J. Rees and R.J. Stoneham, Dordrecht: Reidel, p. 267.

Branch, E. 1984, in: Eleventh Texas Symp. Relativ. Astrophys., ed. D.S. Evans, Ann. New York Acad. Sci. 422, 186.

Deming, D., Rust, B.W., and Olson, E. 1973, Publ. Astron. Soc. Pacific 85, 321.

de Vaucouleurs, G. 1984, preprint.

de Vaucouleurs, G., and Buta, R. 1981, Publ. Astron. Soc. Pacific 93, 294.

Elias, J.H., Frogel, J.A., Hackwell, J.A., and Persson, S.E. 1981, Astrophys. J. Letters 251, L13.

Flin, P., Karpowicz, M., Murawski, W., and Rudnicki, K. 1979, Acta Cosmologica 8, 5.

Green, D.A. 1984, Monthly Not. Roy. Astron. Soc. 209, 449.

Iye, M., and Richter, O.-G. 1984, ESO Preprint No. 347.

Jones, K.G. 1976, J. Hist. Astron. 7, 27.

Kowal, C.T. 1968, Astron. J. 73, 1021.

Kraan-Korteweg, R.C. 1985, to be published.

Minkowski, R. 1964, Ann. Rev. Astron. Astrophys. 2, 247.

Oke, J.B., and Searle, L. 1974, Ann. Rev. Astron. Astrophys. 12, 315.

Panagia, N. 1982, Third European IUE Conf., eds. E. Rolfe, A. Heck, and B. Battrick p. 31.

Pskovskii, Yu.P. 1977, Astron. Zh. 54, 1188.

Rust, B.W. 1974, Thesis, Univ. Illinois.

Sandage, A., and Tammann, G.A. 1982, Astrophys.J. 256, 339.

Sandage, A., and Tammann, G.A. 1984, in: Large-Scale Structure of the Universe, Cosmology and Fundamental Physics, eds. G.Setti and L.Van Hove, p.127.

Schwarz, U.J., Arnal, E.M., and Goss, W.M. 1980, Monthly Not.Roy.Astron.Soc. 192, 67 P.

Searle, L. 1983, Ann.Rep.Dir.Mt.Wilson and Las Campanas Obs.1982-1983, p.622.

Sutherland, P.G., and Wheeler, J.C. 1983, preprint.

Tammann, G.A. 1978, Mem.Soc.Astron.Italiana 49, 315.

Tammann, G.A. 1979, in: ESA/ESO Workshop on Astronomical Uses of the Space Telescope, eds.F.Macchetto, F.Pacini, and M.Tarenghi, p.329.

Tammann, G.A. 1982, in: Supernovae: A Survey of Current Research, eds. M.J. Rees and R.J.Stoneham, Dordrecht: Reidel, p.371.

RADIUS AND ABSOLUTE MAGNITUDE OF A CEPHEID VARIABLE

A. J. Wesselink
Yale University

I. Introduction

Cepheid variables can now be used with confidence for the determination of the distances of all stellar systems in which they occur and can be seen, as two independent methods agree on their absolute magnitudes. Unfortunately, they are too faint to be seen beyond the local group of galaxies and hence are of no use for the distances of galaxies that are more remote. On the other hand, Supernovae are up to 100,000 times brighter than Cepheids and can, consequently, be seen and recognized in stellar systems that are far beyond the local group. They would be ideal distance indicators if only their luminosities were more accurately known. We are here assembled with the desire that we shall improve on this situation and use a known bright luminosity for the determination of distances far greater than is possible with Cepheids. In the meantime, I would like to discuss our present knowledge of the diameters and luminosities of the Cepheid variable. Before embarking on my subject, I should like to remind this audience that at least three of the great names connected with our subject were living and working in this area of our conference:

Henriette Leavitt—The discovery of the Period-Luminosity relation.
Harlow Shapley—The hypothesis that Cepheids are pulsating spheres.
Cecilia Payne-Gaposchkin—Contributions to variables of all sorts including Novae and Supernovae.

II. Baade's Proposal of Measuring Diameters of Cepheids and their Luminosities from Photometry and Radial Velocities

Harlow Shapley (1914) introduced the hypothesis that Cepheids are radially pulsating variables in which the radius and the surface-brightness vary continuously with the period of the light variation. The periodic change in radius is shown by the variable radial velocity. The variation of the surface-brightness is indicated by the continuous changes in color and spectrum. The combined effect of these changes produces the variations in light that are directly observed. Accordingly we have:

$$l = r^2\, S \tag{1}$$

where l is the brightness, r is the radius, and S represents the surface-brightness of the star. We shall refer to equation (1) as Shapley's equation. By taking logarithms and multiplying with $-2\frac{1}{2}$, we obtain the magnitude form of equation (1):

$$m = s - 5 \log r\,. \tag{2}$$

In this formula, m is the magnitude, s is the magnitude of the surface-brightness, and r is the radius.

It was Baade (1926) who proposed that one should try to determine each of the two terms on the right hand side of equation (2) by making use of the light curve, color curve, and radial-velocity curve. He realized that such an effort, if successful, would lead to a test of the pulsation theory as proposed by Shapley and to determinations of the radius and the absolute magnitude. Baade never tried his own suggestion, probably realizing that the data were not accurate enough to be successful. Actual attempts were made to resolve equation (2) by others (e.g., Bottlinger 1928), in the late twenties, but failed. The data were simply not good enough at the time. Furthermore, one tried to determine s in bolometric light. The conversion of the color-index into bolometric correction and effective temperature could not be done with the needed precision. We might discuss in this conference whether such a transformation would be feasible even today.

The early forties saw renewed activity in the field. W. Becker (1940) and W. Strohmeier, with new data, found radii and absolute magnitudes for a number of Cepheids. Their luminosities were in fair agreement with Shapley's P-L relation. Another astronomer working on our subject at the time was A. van Hoof (1943). His results did not become known until after World War II. By 1944, I had amassed enough observations to construct an accurate photovisual light curve of δ Cephei, which combined with the blue photoelectric light curves of Smart (1935) and Guthnick (1921) yielded a color curve of high precision.

I used these photometric data and the radial velocities by Moore (1913) and Jacobsen (1935) for my own analysis of the Baade proposal (Wesselink 1946a, b; see also Wesselink 1969). Soon thereafter, I made a second attempt by replacing the photometry by the very accurate six-color photoelectric photometry by Stebbins (1945). I proceeded as follows. We may write Shapley's equation (2):

$$m = s(c) - 5 \log (r_o + D) \tag{3}$$

where s is the surface-brightness, c the color index, and r_o the minimum radius. D is the displacement of the surface above the minimum position. For further information on the evaluation of D, we refer to the Appendix. We will proceed considering D to be known. The plan is to determine $s(c)$ in the visual system. We thus avoid the difficult problem to convert color indexes into reliable bolometric corrections and effective temperatures.

In our first attempt, the relation between s and c will be taken to be linear:

$$s = bc + \text{const} . \tag{4}$$

A further simplification is possible by realizing that D is relatively small with respect to r_o. Hence, developing the logarithm in (2), the equation of condition is obtained with the color index in the right-hand side:

$$m \cdot x + D \cdot y + z = c\,. \tag{5}$$

Here, x, y, and z are to be determined by least squares; z is an unimportant constant, $x = \frac{1}{b}$, and $y = \frac{5 \log e}{b r_o}$, from which r_o is derived. The standard error of y and r_o will be found in the conventional way. It is explained in the Appendix why the conditional equations were arranged with the color index in the right-hand side.

III. Second Solution, Not Assuming a Linear Relation Between s and c

An improved solution of our problem in which the relation between s and c is not assumed to be linear but of arbitrary form proceeds as follows:

If in equation (3) an arbitrary value for r_o is assumed, the term $5 \log (r_o + D)$ can be calculated. We then calculate s by adding that term to m. According to (3), the surface-brightness s is obtained. The surface-brigntness s is then plotted against c throughout a cycle of the Cepheid variable. The result will be a closed curve. The procedure is then repeated with different values for r_o until the closed curve condenses into a unique relation. The value of r_o for which this happens is the value of the radius that we seek.

The standard error of r_o is most easily obtained with a method first developed by Pannekoek and van Dien (1937). We refer to the literature for the details of this method. It will again be necessary to evaluate the sum of the squares of the residuals from the $s(c)$ curve in the direction of the color index c.

IV. Results of the Computation for δ Cephei and Comparison with the Absolute Magnitude as Given by the Period-Luminosity Law

I found $r_o(\delta \text{ Cephei}) = 48\,r$ (sun) with a standard error of $1.7\,r$(sun) as my (1945) value for the radius. The mean intrinsic color is $0\overset{m}{.}04$ more blue than the sun, resulting in a surface brightness $0\overset{m}{.}15$ brighter than the sun. Combining the differences of surface-brightness and radius, we find δ Cephei $8\overset{m}{.}55$ brighter than the sun. With an absolute magnitude for the sun of $+4\overset{m}{.}85$ we obtain:

$$M_V\ (\delta \text{ Cephei}) = -3.70\,.$$

At the time, Shapley's period-luminosity relation was generally accepted and gave M_V (δ Cephei) $= -2.20$. The difference of $1\overset{m}{.}5$ between my pulsation result and Shapley's P-L relation was considered real and most disturbing by me. I questioned my analysis rather than Shapley's P-L relation.

However, seven years later (1952), Baade produced evidence that the distance scale had to be enlarged approximately by a factor 2. This implied a correction to Shapley's P-L relation of $-5 \log 2 = -1\overset{m}{.}5$. This removed in one blow the discrepancy between my pulsation absolute magnitude and the P-L relation and restored my confidence in the type of analysis outlined in this paper.

Appendix

I. Evaluation of the Displacement D

The measured radial velocity is the integrated result over the stellar disk of the component of the expansion velocity in the direction of the observer. The factor required to convert radial velocities into expansion velocities depends on the amount of limb darkening. It is $-\frac{3}{2}$ for a uniformly bright disk, $-\frac{4}{3}$ for a completely darkened disk. If v is the expansion velocity with respect to the star's center, the displacement at any time t is given by $D = \int_0^t v\,dt$; where at $t = 0$, the radius is minimal.

II. Remark on the Form of the Equation of Condition in the Least-Squares Solution

In any least-squares solution, the coefficients of the unknowns in the conditional equations are considered without error. This is the part of the equations that usually occurs at the left of the equality sign. The variable containing the error then occurs on the right side. In our case, the position of the variables requires careful consideration since all three variables m, D, and c are subject to observational error. It can be shown that in our case the form with c adjustable, i.e., on the right side, is the correct one. Only in this form is it guaranteed that the residuals of the solution are accidental errors almost entirely due to c itself, the errors in m and D being negligible.

References

Baade, W. 1926, *A. N.* **228**, 359.

Becker, W. 1940, *Mitt. Potsdam* **4**.

Bottlinger, K. 1928, *A. N.* **232**, 3.

Guthnick, P. 1921, *Jubileum A. N.* **10**.

Jacobsen, T. S. 1935, *Lick Obs. Bull.* **12**, 243.

Moore, J. H. 1913, *Lick Obs. Bull.* **7**, 153.

Pannekoek, A. and van Dien, E. 1937, *B. A. N.* **8**, 141.

Shapley, H. 1914, *Ap. J.* **40**, 448.

Smart, W. M. 1935, *M. N. R. A. S.* **95**, 644.

Stebbins, J. 1945, *Ap. J.* **101**, 47.

van Hoof, A. 1943, *Kon. Vlaamsche Akad.* **5**.

Wesselink, A. J. 1946a, *B. A. N.* **10**, 83.

Wesselink, A. J. 1946b, *B. A. N.* **10**, 251.

Wesselink, A. J. 1969, *M. N. R. A. S.* **144**, 297.

ASTROPHYSICAL DISTANCES TO TYPE II SUPERNOVAE

Robert P. Kirshner
Department of Astronomy
The University of Michigan

I. INTRODUCTION

New techniques in astronomy often lead to new insights into astronomical objects. The advent of linear detectors in astronomical spectroscopy has created a discrete leap in our understanding of supernovae. One of the most intriguing possibilities that stems from this improvement is a way to measure distances to extragalactic supernovae by studying their expanding atmospheres from their continua and lines. In particular, the Type II supernovae (SN II), which are found in the spiral arms of spiral galaxies, have a smooth, well-defined continuum and easily identified Balmer lines that lend themselves to this technique.

The ability to measure distances to individual objects means that the supernovae need not be standard candles, and the problems with sampling bias and evolutionary corrections that plague other methods may not dominate here. Since the distances to bright SN II are tens of megaparsecs, this method skips all the treacherous rungs on the cosmic distance ladder from the parallax of the Hyades to the calibration of the Cepheids and can give distances well beyond the Virgo cluster. While astrophysical distances to supernovae have uncertainties, the developing physical picture of supernova envelopes holds out the hope that the error can be reduced as understanding increases.

The introduction of another new technique, VLBI of radio emission from supernovae (as described by Bartel in this volume), promises a complementary, but essentially similar way to gauge the distances to supernovae. Both require a deeper understanding of supernova explosions to reach maturity: the optical method needs more refined model atmospheres and the radio work needs a more thorough understanding of the way supernova ejecta generate relativistic particles (see the contribution by Chevalier). The two may be closely linked in that the material which forms the optical continuum at early times may go on to generate the radio emission at late times. If that is the case, then the proper interpretation of the radio data hinges on a correct understanding of the optical line profiles.

II. TECHNIQUES

The introduction of linear detectors, such as Oke's Multichannel Spectrometer , Wampler's Image Dissector Scanner, and Shectman's Intensified Reticon allowed accurate flux measurements at moderate resolution over a wide wavelength span, and provided an effective way to subtract both night sky emission and the possible contamination from a galaxy in the spectrum of a supernova. The happy coincidence of new detectors, several bright supernovae of both Type I and Type II, and my own arrival at Caltech as a graduate student gave me a chance to explore the potential of this new type of data for improving the understanding of supernovae (Kirshner et. al. 1973, Oke and Searle 1975). New results on the continuum shape, and line shapes led to a new way to find distances for Type II supernovae (Kirshner and Kwan 1974, Kirshner and Kwan 1975, Kirshner 1977).

Some key results are indicated in Figure 1, which shows a pair of multichannel scans of the supernova 1973r in NGC 3627, made 26 and 128 days after maximum light. In both scans, most of the energy is carried by the continuum, and good fits to it are provided by blackbodies near 5000 K, as hinted by earlier work from broadband UBV photometry (Arp 1961, Psovskii 1969). In general, the evolution of the continuum temperature is illustrated in Figure 2, which shows a temperature at maximum in the range 12000-20000K followed by rapid cooling to about 5000 - 6000 K in the thirty days after maximum light. As Figure 1 shows, the continuum temperature remains roughly constant for at least another 100 days.

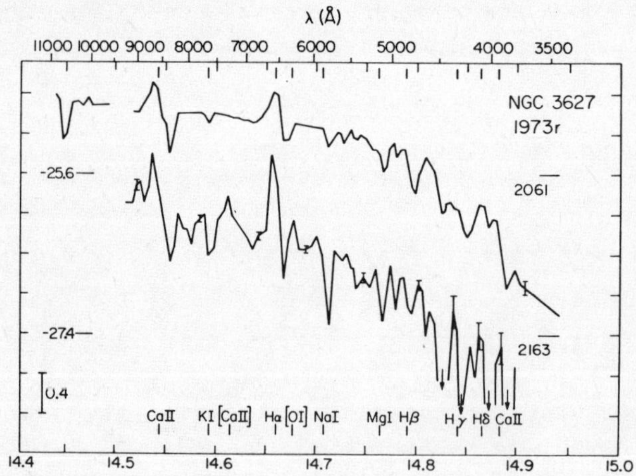

Figure 1. Multichannel scans of the SN II 1973 R in NGC 3627. The scans are at 26 and 128 days past maximum light. The continuum has a temperature near 5000 K.

The temperature comes from the shape of the continuum, but these measurements also include absolute calibration that permits the absolute flux density, f_ν, to be measured to an accuracy of a few percent. Though it is too simple, the assumption of a spherical blackbody allows the combination of temperature, T, and f_ν to yield the angular size :

$$\Theta = \frac{R}{D} = \left(\frac{\pi B_\nu(T)}{f_\nu}\right)^{1/2} \quad (1)$$

Of course, we attribute the change in angular size to increasing radius, R, rather than changing distance, D despite substantial secular changes in the extragalactic distance scale. Figure 3 shows that just as you might expect in a violent stellar explosion, the radius of the photosphere increases with time during the first month. Even though the supernova is decreasing in luminosity through the first month after maximum, the decline in temperature is enough to make the inferred radius increase.

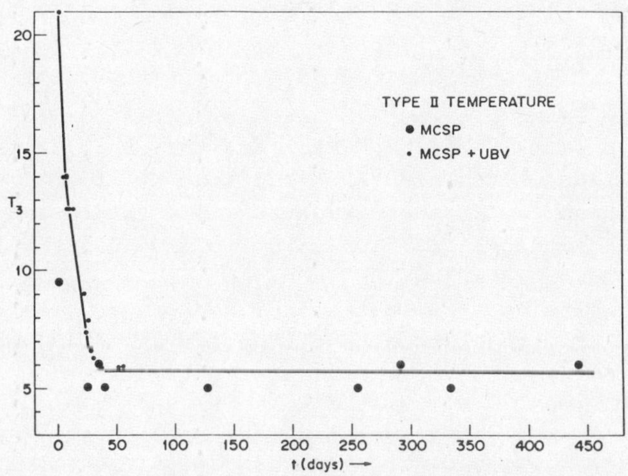

Figure 2. Temperature evolution for a SN II.

Of course, a blackbody is too strong an assumption to use without further investigation, and there are other assumptions involved in the method described here. To help the skeptics keep score, Table 1 lists the principal assumptions, some observations that can be used to check them, and the next step forward in refinement. For the blackbody assumption, comprehensive observations from the UV to infrared are desirable to look for deviations from the Planck law. Model atmospheres, such as those of Hershkowitz and Wagoner (1985) are the next step.

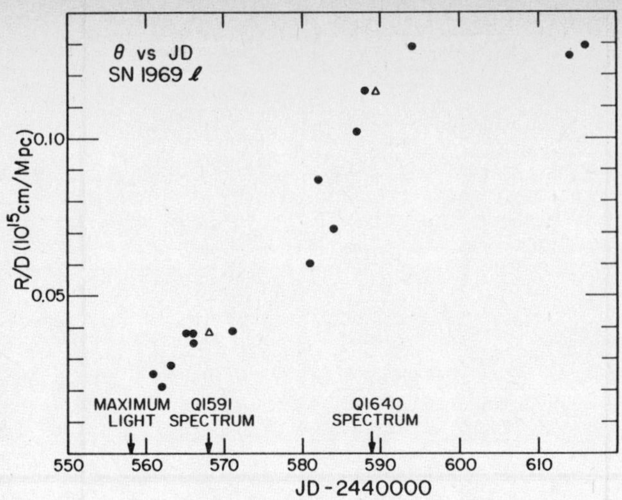

Figure 3. Angular size of the photosphere of a SN II.

On inspecting the photometric scans of supernovae deep in the second sub-basement of Robinson Lab, within earshot of Zwicky himself, Leonard Searle pointed out that the angular expansion can be compared to the rate of expansion inferred from absorption lines to give the distance to the supernova. This was applied by Branch and Patchett (1973) to SN I's and by Kirshner and Kwan (1974) to SN II.

To make some progress, it is necessary to make some assumptions, but they can be checked later against other evidence. If the envelope undergoes free expansion after a brief period of acceleration when the emerging shock hits the atmosphere of the star, then material with velocity v will be at a distance R at a time t given by:

(2) $$R = v\,(t - t_o) + R_o$$

where the initial radius is R_o at the moment t_o. Now R_o and t_o are not accessible to observation so the sensible thing to do is to use two or more observations of the flux, temperature, and velocity separated by a sufficient span of time. Then

(3) $$R(t_1) = v_1(t-t_0) + R_0$$

(4) $$R(t_2) = v_2(t-t_0) + R_0$$

using (3) to eliminate t , we get

(5) $$R(t_2) = v_2(t_2-t_1) + \frac{v_2}{v_1}(R_1-R_0) + R_0$$

and eventually a simple equation for the distance in terms of observables and R_0, which can be constrained from observations and rendered unimportant by well-spaced data:

$$D = \frac{v_2(t_2-t_1) + R_0\left(1-\frac{v_2}{v_1}\right)}{\Theta_2 - \frac{v_2}{v_1}\Theta_1} \tag{6}$$

TABLE 1
INGLORIOUS ASSUMPTIONS

ASSUMPTION	OBSERVATIONAL TEST	IMPROVEMENT
Supernova continua are blackbodies	Energy distribution from IR to UV	Model atmospheres: Hershkowitz & Wagoner (1985)
Supernovae are spherical	Polarization (see McCall in this volume)	Not needed?
R_0 small	Pre-explosion limits (Thompson 1982)	Not needed for well-spaced observations
$V \propto R$	Evolution of line shapes	Not needed
Density decreases with R	Detailed fit to line shapes (Branch 1980)	Not needed
Velocity of dense shell gives radio expansion	Simultaneous radio and optical data: late time spectra.	More theory of radio emission mechanism

It is worth noting that this approach does not assume that the rate of change of the photosphere's radius is the same as the velocity of the gas that is instantaneously at the photosphere. In general, as the supernova envelope expands, the supernova photosphere will recede in lagrangian coordinates so $\dot{R}$ will be smaller than v. It is also worth noting that this approach does not require that the observed velocity of the material at the photosphere remain constant. For observations which are spaced so that $v_1 \cong v_2$, the effect of an unknown R_0 is small. For any reasonable guess at the initial radius the effect is not important. Even better, observations of the galaxy before the supernova can give limits on the luminosity of the pre-supernova star, and provide contraints on R_0 (Thompson 1982).

So the problem in determining distances to SN II's becomes one of determining the velocity of the material at the photosphere by interpreting the shapes of the absorption lines. Superposed on the continuum, SN II's have strong absorption lines formed in the gas between us and the photosphere. The lines are similar in shape to those seen in stars with stellar winds, the prototype of which is P Cygni. In this case, although the velocities are much higher, the radiative transfer problem is very similar, and the characteristic P Cygni lines with blue-shifted absorption provide significant clues to the velocity and structure of the supernova atmosphere. Figure 4 shows observations obtained with good resolution and good signal-to-noise that illustrate the typical shape of the Balmer lines in SN II spectra. They show a broad absorption trough extending far to the blue of each Balmer line and an asymmetric emission centered near zero velocity. In the upper spectrum, Hα absorption extends from -6000 to -12000 km/sec.

Figure 5 illustrates schematically how lines are formed by scattering in an expanding atmosphere. If we assume that distance is proportional to velocity as in equation (1), then the locus of the atoms which absorb a particular velocity is just the disk perpendicular to the line of sight illustrated in Figure 5. So the problem of computing the line profile is to calculate the fate of photospheric photons travelling out through this scattering atmosphere, as seen by an observer, as discussed in some detail by Branch (1980) and Branch et. al (1981). The most important qualitative result is that the velocity of the gas at the photosphere is given by the red edge of the absorption minimum.

This result can be understood by looking at Fig. 5. The most blueshifted absorption comes from material that is far from the photosphere. As long as the density of scatterers decreases with increasing velocity, there will be relatively little absorption at the highest velocity. Inspection of Figure 4 shows that in fact the absorption does taper off toward the blue edge of the profile, a fact which requires a declining density.

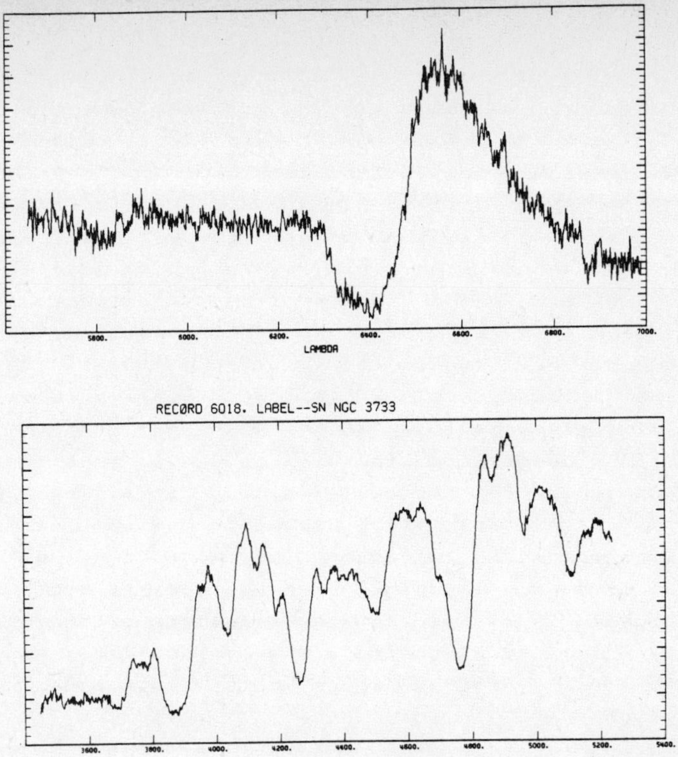

Figure 4. Spectra of a SN II showing P Cygni line profiles.

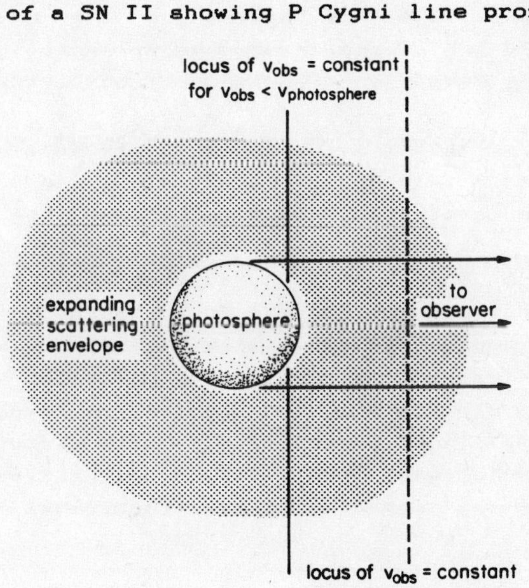

Figure 5. The expanding atmosphere of a SN II where scattering occurs. The loci of constant observed blueshift are shown for the case where v is proportional to R.

In any event, as the atmosphere expands, its motion will create an atmosphere in which density drops off at least as fast as $1/r^2$. More to the point, every calculation of the break-out of a shock from a stellar atmosphere indicates that the initial density distribution is expected to be a steep function of velocity (Grasberg et. al. 1970). As we consider disks which are closer to the photosphere, we expect more absorption until we reach the disk which is just tangent to the photosphere. Beyond that point, for yet lower blueshift, the area of the photosphere from which absorption is possible begins to shrink, so we expect less depletion of the photospheric continuum. All of this means that the maximum absorption will occur right at the photospheric velocity.

The actual situation is a bit more complex, because of the possibility of large optical depths and the effects of emission scattered into the observer's line-of-sight, but modelling shows that the general conclusion is correct. The way to find the photospheric velocity is to seek the absorption minimum, or for line profiles with flat absorption troughs, the red edge of the absorption. The weakest lines have the least complications from the optical depth effects, so the way to proceed is to use the higher members of the Balmer series, such as $H\beta$, $H\gamma$, and $H\delta$ to trace the evolution of the photospheric velocity with time. Figure 6 shows measurements made according to this prescription.

Two points are important. First is that the photospheric velocity changes rapidly at the beginning. That means that the outer layers of the expanding atmosphere become transparent as the envelope expands and we see deeper into the mass of the star. It also means that velocities need to be evaluated at several times in the first month to do the distance estimate. The second is that the decrease in velocity stops its rapid change at about age 40 days and declines very slowly for at least another 200 days. This requires that a very large fraction of the envelope mass is constrained to a small range of velocity of order 5000 km/sec. Put another way, a shell of material has been ejected that continues to form absorption lines even after the continuum has receded to material with very low average velocities of a few hundred kilometers per second. That same dense shell is presumably responsible for the fact that the photospheric angular size increases so nearly linearly with time. While equation (6) does not require a linear increase in θ , the opportunities for error are diminished by this good behavior of real supernovae.

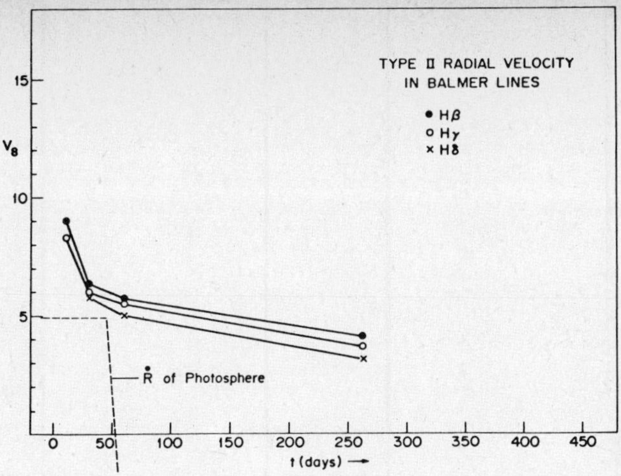

Figure 6. Velocities in a SN II atmosphere measured from the Balmer lines.

At times past 50 days, the photosphere begins to shrink rapidly and the overall picture does not correspond to Figure 5 . In the broadband light curves for SN II, this epoch marks a rapid decline. What is really happening is that the dense shell has become transparent in the continuum. There is, however, material at lower velocities, because the continuum does not vanish, as shown in Figure 2. The material seen at such late times has an average expansion velocity of only a few hundred kilometers per second. This material continues to obscure the central object (if there is one) and the possible exotic products of nucleosynthesis in SN II's. The dense shell of material at about 5000 km/sec may be the relevant part of the supernova ejecta for the evolution of radio emission and its velocity is of special importance for interpreting the observations of radio angular diameters. Another way of estimating the velocity of the dense shell may be to look at the width of the emission at Hα at late times.

III. DISTANCES

Actually applying this method to derive some distances to supernovae has been a very slow job. The obstacles are not so much technical as organizational. In general, observatory schedules are not constructed in a way that makes it easy to get several observations of a suddenly discovered object during the first month, all with the same instrument. The places where this can happen either have one instrument, which happens to be suitable, that is almost always on the telescope or they have very strong interest in supernovae and are willing to upset all the other observers to get the data.

Other practical issues also intervene. The observations need good weather so the flux calibrations can be accurate, good seeing so the light goes down the aperture, and the object has to stay above the horizon long enough for high quality data to be obtained.

All excuses aside, Table 2 gives the results for four SN II's with extensive spectrophotometry for which the distances have been derived. The detailed data for SN 1979c and 1980k will be published elsewhere, but it is all spectrophotometry obtained with the McGraw-Hill Observatory 1.3 meter telescope using an intensified reticon spectrograph. The temperatures and velocities have all been determined from the scans alone. In the case of M100, an independent analysis is available from the Texas group (Branch et. al. 1981). There are results for a few other SN II's from a more model-dependent approach by Schurmann, Arnett, and Falk (1979), but they are not included here even though they are generally consistent.

TABLE 2
ASTROPHYSICAL DISTANCES TO SN II

SUPERNOVA	GALAXY	V SHELL (KM/SEC)	D (MPC)	ERROR	REFERENCE
1969 L	NGC 1058	4600	12	4	Kirshner and Kwan 1974
1970 G	M 101	4300	7	2	Kirshner and Kwan 1974
1979 C	M 100	9000	21	4	This work
			23	3	Branch et. al. 1981
1980 K	NGC 6946	4600	12	4	This work

The galaxies that have SN II's with astrophysical distances are all of special interest in the overall program to determine the extragalactic distance scale (see Sandage & Tammann, this vol.). They are the luminous, relatively nearby spirals in which other distance indicators may be found. While the current data should be regarded with caution, all the cases to date are consistent with a distance scale that corresponds to a Hubble constant near 60 km/sec/Mpc.

The supernova in M100 is of outstanding interest because it is the biggest spiral galaxy in the Virgo cluster. Even though that supernova was unusual in many ways, the fact that this method does not require SN II's to be standard candles allows a relatively secure distance to be assigned.

For convenience, Table 2 also lists the velocity associated with the dense shell. It seems plausible that this velocity might be the relevant one for thinking about the interaction of the supernova ejecta with the surround to produce radio emission.

There is no doubt that an intensified effort to measure the distances to SN II could yield a dozen well-observed cases in the next 5 years. This will require improved searches, which are already being initiated, and a strong observational effort which requires broad cooperation based on an understanding of the value of these observations. Observations with the Space Telescope could be of special value in providing good UV data to check the overall energy distribution, in providing estimates of the reddening, in avoiding confusion with background, and in photometric measures at late times.

Observations of SN II at early times to establish the behavior of the photosphere and at late times to measure the width of the emission lines may be of particular importance in interpreting the VLBI data. A deeper understanding of the connection between the material observed optically and the generation of the radio emission needs to be developed, although the suggestions of Chevalier (this volume) provide a good place to start.

Research on supernovae at the University of Michigan is supported by the National Science Foundation.

REFERENCES

Arp, H.C. 1961, Ap.J.,**133**,883.

Branch, D. 1980, Proc. Workshop on Atom. Phys. and Spectrsc. Supernovae Spectra, ed. R.E. Meyerott (New York: AIP)

Branch, D. and Patchett, B. 1973, M.N.R.A.S., 161,71.

Branch, D., Falk, S.W., McCall, M.L., Rybski, P., Uomoto, A.K., and Wills, B.J. 1981, Ap.J.,**244**,780.

Grasberg, E.K., Immshennik, V.S., and Nadezhin, D.K. 1971, Astr. Space Sci.,**10**,28.

Hershkowitz, S. and Wagoner, R. 1985, Abstract-- Tucson AAS.

Kirshner, R.P. and Kwan, J. 1974, Ap.J.,193,27.

Kirshner, R.P. and Kwan, J. 1975, Ap.J.,197,415.

Kirshner, R.P., Oke, J.B., Penston,M.V., and Searle, L. 1973, Ap.J.,**185**,303.

Oke, J.B. 1969, Pub. A.S.P.,**81**,11.

Oke, J.B. and Searle, L. 1975, Ann. Rev. Astron. Astrophys.,**12**,315.

Psovskii, Yu.P. 1969, Soviet Astr. --AJ,12,750.

Schurmann, S.R., Arnett, W.D., and Falk, S.W. 1979, Ap.J.,230,11.

Thompson, L.A. 1982, Ap.J. (Letters),**257**,L63.

MODEL ATMOSPHERES FOR TYPE I SUPERNOVAE

Robert Harkness
Astronomy Department
University of Texas at Austin

In recent years supernova explosions have become a promising new tool for obtaining accurate distance determinations on a cosmologically significant scale. Type I supernovae probably offer the best means of determining not only the Hubble constant, but also the cosmological deceleration parameter via observations of very distant events (up to z = 0.3) using the Space Telescope (Wagoner, 1977,1979). Type I's are particularly attractive because of their higher peak luminosity. Also, the remarkable similarity in their light curves and spectral development leads one to suspect that Type I supernovae may be critical events (such as a white dwarf accreting matter and exceeding the Chandrasekhar limit) and hence independent of any cosmological evolutionary influences. There are essentially two approaches to using supernovae as distance indicators.

Firstly, Type I's may be useful as "standard candles". However, there seem to be a range of subclasses of Type I events; for example, those with "slow" light curves and high peak luminosities and vice versa (Branch 1981,1982). Also, there are now at least two well observed "anomalous" Type I's (M83 in July 1983 and NGC0991 in August 1984, see Wheeler, this volume) so it seems that detailed spectroscopic observations extending from the earliest possible time after the explosion may be necessary to confirm "subtype" and hence peak luminosity even in the "standard candle" approach.

Most attention, however, has been focussed upon the use of the Baade-

Wesselink method as a direct means of distance determination. This method has been used with supernovae of both types (Branch and Patchett 1973, Kirshner and Kwan 1974), resulting in plausible distance estimates, but there remain doubts about the application of this simple approach to the complex conditions prevailing in a supernova atmosphere.

A thorough understanding of supernova atmospheres is needed now to determine the corrections which may be necessary before one can place much confidence in the Baade-Wesselink distance estimates. This discussion will be limited to models of Type I supernovae and in particular the consequences of the detonating or deflagrating white dwarf mechanism. The term "supernova atmosphere" may be seen to be misleading because there is only a formal connection with normal stellar atmospheres. The "atmosphere" of a Type I supernova rapidly becomes most if not all of the mass in an exploding white dwarf model.

The Baade-Wesselink method requires measurements of "photospheric" velocity and temperature and flux at some wavelength (preferably in the continuum) at two or more times. Classically, the effective temperature is obtained from the slope of the optical continuum and the photospheric velocity is taken to be equal to the velocity of the blueshifted absorption minimum in lines which are assumed to be P-Cygni profiles due to resonance scattering above a sharp, blackbody photosphere. If the photosphere does radiate as a sharp spherical blackbody the angular size of the supernova at any time is simply the square root of the ratio of the observed flux to the blackbody flux which would be radiated towards an observer from a blackbody at the measured effective temperature. If the radius of the photosphere can be represented as $R = vt + \text{constant}$, the distance to the supernova is obtained directly from the definition of its angular size. Thus, the validity of the method rests on the assumptions that the photosphere is geometrically sharp, radiates as a blackbody and that the observed line velocity coincides with the expansion velocity of this photosphere.

What, then, are the potential pitfalls? In the spectra of Type Is there are some unblended features which can be identified with reasonable certainty (especially Ca II and Si II) but it is nearly impossible to identify the continuum in the optical region. These lines may well be formed above the radius of unit continuum optical depth and furthermore, may extend over a considerable range in velocity, particularly if a strong abundance stratification exists. In this case the expansion velocity will be overestimated. Even so, the spectral synthesis models of Branch (1984 and this volume) and the atmosphere models described here show that it is unlikely that the estimated velocities can be in error by a large factor (probably 0.8 to 1.5), because the predicted line profiles would produce an unacceptable fit to the observed spectrum.

The most critical parameter is the effective temperature. This is estimated from the colour temperature, but there are several physical effects which can lead to large over- and under-estimates of the effective temperature required in the blackbody approximation. Exploding white dwarf models tend to result in a density profile which becomes an increasingly steep function of radius, but even before maximum light the atmosphere has considerable "extension" in the sense that a plane-parallel approximation is inadequate because the radius of unit optical depth becomes extremely frequency dependent. The importance of atmospheric extension increases as the density profile in the "photospheric" region flattens out. By the time the photosphere has receded to the half mass point this effect is large. Atmospheric extension results in a continuum which is "flatter" than a plane-parallel atmosphere at the same characteristic temperature and hence optical colour temperatures are an under-estimate of the overall effective temperature. Thus significant atmospheric extension results in a systematic (and time dependent) under-estimate of the luminosity. Probably the most serious difficulties occur if the continuum opacity is dominated by electron scattering. Wagoner (1981,1984) has shown that for Type II supernovae this is likely to be true. For the Type I atmospheres presented here electron scattering is the main component of continuous opacity at optical wavelengths to large optical depths. The continuous absorption arising from excited states is difficult to estimate for the compositions encountered because the relevant data is not available at present. The main consequence for the Baade-Wesselink method is that the characteristic temperature of the optical continuum is the temperature at the radius where the radiation field is thermalised. Thus the observer sees a diluted but "higher temperature" blackbody. If the radius at which the radiation is thermalised is widely separated from the radius at which it is last scattered the corrections will be very significant. Obviously, the effect is to over-estimate the luminosity.

To determine the overall effects one needs to construct a detailed model atmosphere. However, the conditions in supernova "atmospheres" are quite unlike those found in conventional atmospheres or stellar winds. The expansion of a supernova rapidly becomes homologous and one has to account for the effects on the radiation field throughout the whole star, which is a very unusual case. Also, the total optical depth to the centre of the supernova diminishes rapidly and is relatively small even at maximum light (i.e "small" means a total optical depth of less than 1000). This can present an enormous problem at later times because of the need to thermalise the source of gamma ray photons arising from the decay of radioactive cobalt which is produced in the "standard" Type I model. Therefore the Type I supernova "atmosphere" is actually a model of the whole star, and the range of physical variables coupled with the need to use a full treatment of the radiative transfer results in a large and complex computation. Details of the radiative transfer codes and their particular application to Type I models will be given elsewhere. The important features are concerned with the solution of the special relativistic radiative transfer equations in spherical geometry using the co-moving form and numerical methods outlined by Mihalas, Kunasz and Hummer (1976) and Mihalas (1980). The models provide for radially dependent elemental abundances and include most available sources of continuous opacity. The computational cost scales with the number of spectral lines included and at present only a few tens of lines can be represented. The ionisation and excitation are currently assumed to be in LTE to keep the calculation manageable although non-equilibrium and non-LTE can be added in principle. The co-moving approach avoids the need to define a frame-dependent "expansion opacity" (Karp et al 1977, Karp 1980), but when the co-moving solution is complete a separate observer frame calculation is needed to obtain the emitted flux. This calculation can also be time-consuming if many spectral lines are necessary. Fortunately, it appears that relatively few very strong lines are enough to account for the observed spectrum because of the large effective bandwidth of each line in the high velocity Type I explosion.

Before the co-moving solution can be started a grid of radius points must be selected on the basis of mean opacities and expected gross properties of the model. With a density profile and elemental abundance profiles from detailed explosion model a co-moving mean opacity can be derived to help in the determination of this radial grid. The outstanding problem concerns the specification of the (co-moving) specific intensity at the inner (core) boundary of the grid. Even at times close to maximum light this boundary may be within the region where the gamma rays are being generated and deposited. If this is the case then one must follow the transport of gamma radiation in a hydrodynamical model. This has not been done consistently for any explosive models. On one hand there are complex explosive models which include nuclear reaction networks (Nomoto et al. 1984) but do not evolve the model beyond a few seconds (and therefore do not treat the radioactive decay and gamma transport) and on the other there are calculations of explosions without detailed nucleosynthesis but which are concerned with the gamma ray deposition and escape until the ejecta are optically thin (Wheeler and Sutherland, 1984). Neither class of models incorporates a mean opacity which is self-consistent with the radially dependent abundances and with the co-moving diffusion of radiation. At present, therefore, the inner boundary condition has to be specified in some ad hoc way which leads to gross atmospheric properties that are consistent with the observed properties. In this sense one needs to know beforehand what constitutes a likely "photospheric" temperature such that the observed ion species will be abundant. Thus the spectral synthesis models of Branch are of tremendous value in helping to limit the permissible range of values for maximum light velocities, temperatures and densities because the cost of the atmosphere calculations described here would prohibit any such broad exploration of these parameters.

Figures 1 and 2 show some typical observer frame spectra emerging from maximum light model atmospheres with a power-law density profile and a density profile from a Nomoto et al. carbon deflagration model, respectively. In both cases the abundances are fully mixed and the "photospheric" velocities and temperatures are 12500 km per second and 11000 K , respectively. The elemental abundances are approximately solar with the exception of calcium and cobalt. These models do not contain any helium or hydrogen (i.e. there are composed entirely of "metals").

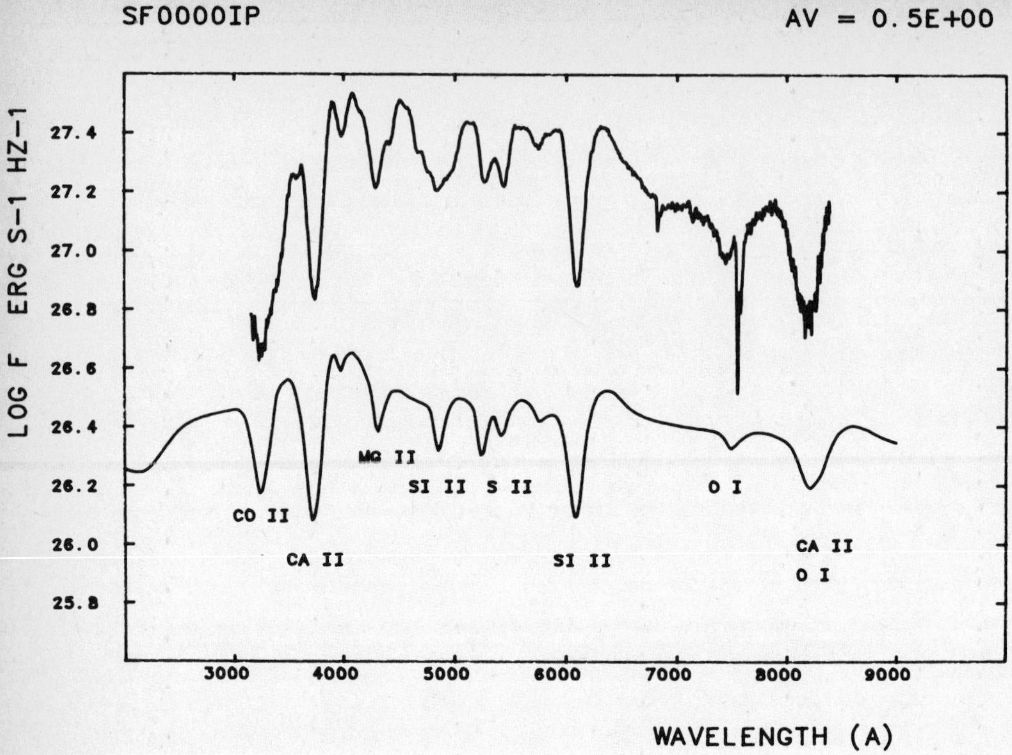

Figure 1. The emergent flux from an atmosphere with a power-law density distribution ($\rho \propto r^{-10}$). Abundances are assumed to be fully mixed, with Co/Fe = 0.1 and Ca/Si = 0.1 $(Ca/Si)_{solar}$. Photospheric velocity and temperature are 12,500 km s^{-1} and 11,000K at radius 10^{15} cm, respectively. The upper curve is the McDonald spectrum of SN1981b at maximum light. A reddening correction of A_ν = 0.5 has been included in the calculated spectrum.

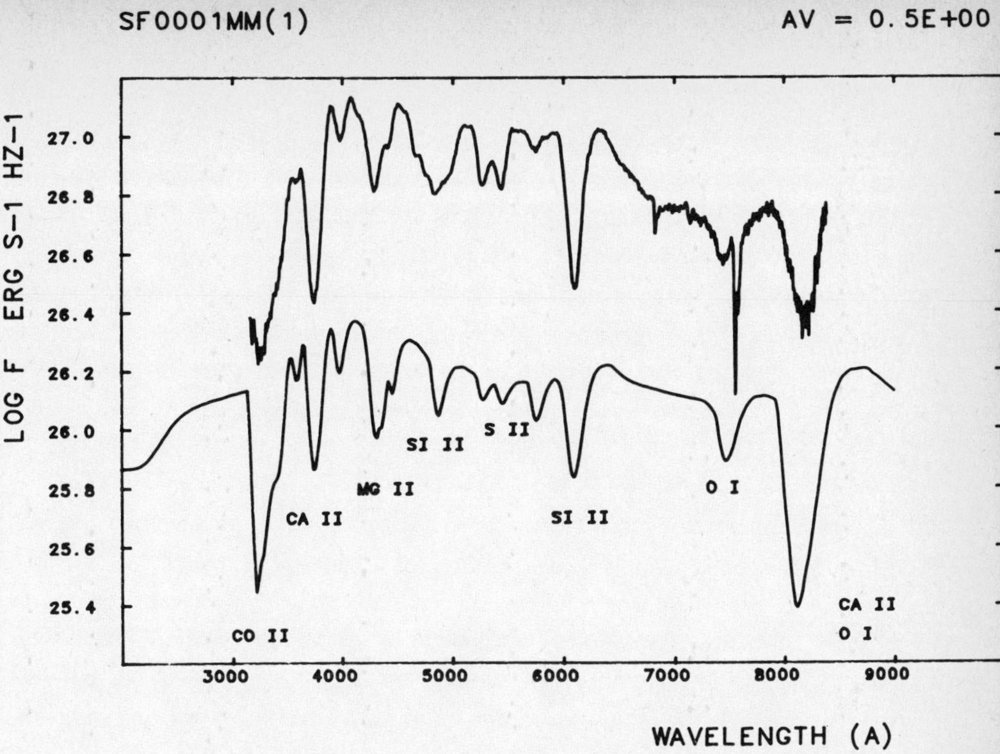

Figure 2. The emergent flux from an atmosphere with the density profile arising from the Nomoto et al. (1984) carbon deflagration model. Abundances, reddening and photospheric properties are the same as in figure 1. Overall the coincidence of the absorption minima is improved and there are some large differences due to the change in atmosphric extension and excitation.

Unfortunately, these particular models CANNOT be used to determine distance estimates by direct comparison with for example, maximum light spectra of supernova 1981b (NGC4536) also shown in the figures because the photospheric radius was inadvertently assigned a value appropriate to a time several days before maximum light. A simple scaling is definitely an unsafe procedure. This does, however, illustrate a crucial point: the Baade-Wesselink method in its basic form is clearly independent of the mechanism of the supernova explosion. Introducing any kind of model atmosphere immediately destroys this attractive aspect. Indeed, construction of a supernova atmosphere itself requires a detailed explosion model to provide the basic structure and any distance estimate will always rely on that fact.

The coincidence of observed and calculated spectral features, both in wavelength (hence velocity and temperature) and relative strength suggests that the favoured explosion model of a deflagrating white dwarf is a very serious contender for the basic Type I mechanism. More comprehensive modelling of the precise observable consequences using the model atmosphere codes mentioned here may be able to discriminate fully between these and alternative explosion mechanisms and also support the use of these objects as distance indicators.

References.

Branch, D., Patchett, B., 1973 MNRAS 161, 71.

Branch, D., Falk, S.W., McCall, M.L., Rybski, P., Uomoto, A.K., Wills, B.J., 1981 Ap.J. 244, 780.

Branch, D., 1981 Ap.J. 248, 1076.

Branch, D., 1982 Ap.J. 258, 35.

Branch, D., 1984, Proceedings of the 11th Texas Symposium on Relativistic Astrophysics.

Karp, A.H., Lasher, G., Chan, K.L., Salpeter, E.E., 1977 Ap.J. 214, 161.

Karp, A.H., 1980, "Supernovae Spectra", Meyerott, R., Gillespie, G.H., eds. American Institute of Physics, New York.

Kirshner, R.P., Kwan, J., 1974 Ap.J. 193, 27.

Mihalas, D., Kunasz, P.B., Hummer, D.G., 1976 Ap.J. 206, 515.

Mihalas, D., 1980 Ap.J. 237, 574.

Nomoto, K., Thielemann, F.-K., Yokoi, K., 1984, preprint.

Wagoner, R.V., 1977 Ap.J. (Letter), 214, L5.

Wagoner, R.V., 1979 Comments on Astrophysics, 8, 121.

Wagoner, R.V., 1981 Ap.J. (Letter), 250, L65.

Wagoner, R.V., 1984 preprint.

Wheeler, J.C., Sutherland, P.G., 1984 Ap.J. 280, 282.

ACCURACY OF MODEL PARAMETERS FROM SPECTROSCOPIC FINE ANALYSES OF SUPERNOVAE

K. Hempe
Hamburger Sternwarte
University of Hamburg
Gojenbergsweg 112
D-2050 Hamburg 80
Federal Republic of Germany

ABSTRACT. A modelgrid of hydrogen Balmer lines has been calculated for supernova atmospheres by the comoving frame technique. The theoretical line profiles have been used for an analysis of SN 1983 in NGC 1448. The comparison of theoretical and observed equivalent widths and line strengths give information on the accuracy of the model parameters as they result from the spectroscopic fine analyses. It has been found that temperature, density, and velocity structure can be well determined, while the photospheric radius is uncertain by a factor of 2. With the knowledge of the age of the supernova it is possible to find a consistent model. The internal accuracy of the model parameters has been found to be of the order of 10 percent.

1. INTRODUCTION

In recent years many synthetic supernovae spectra have been published by several authors (Branch et. al. 1981, 1982, 1983; Axelrod 1980, 1981; Fransson 1984; Hempe 1981, 1983). The comparison of theoretical spectra with the observed spectra show that the general agreement is very well while there are still some severe discrepancies in certain features. However, there has been no study of the internal accuracy or unambiguity of the models which have been used for the calculation of synthetic spectra. The accuracy of the supernova model parameters is especially important for distance determinations. A very powerful tool for distance determinations is the Baade-Wesselink method (Baade, 1926; Wesselink, 1949) in a form modified by Branch et al. (1981). In this method the angular diameter of the supernova is determined by photometric quantities, e.g. the temperature and the visual magnitude of the SN. On the other hand the angular diameter of the supernova can be determined by the expansion velocity of the photosphere, the time since explosion, and the distance of the supernova. Equivalent to these parameters are the radius of the photosphere and the distance of the supernova. Supernova atmosphere models give information on the velocity at the location of the photosphere, the photospheric radius, the temperature, and the density structure.

Here we use a model grid of Balmer lines calculated for supernovae (SNe) type II for an analysis of SN 1983 in NGC 1448. This supernova has been observed by Richter on October 28, 1983 with the IDS at the ESO 1.5 m telescope at La Silla. The observations were made 22 days after discovery of this SN II by Evans and McLean. The calculated model grid can be described by 4 parameters: the velocity at the photosphere v_o, the temperature T in the expanding atmosphere, the photospheric radius R_*, and a density exponent n. This model grid has been used for the calculation of the hydrogen Balmer lines. The line transfer problem has been solved by the comoving frame technique developed by Mihalas, Kunasz and Hummer (1975). From the comparison of the theoretical and observed line profiles we determine a best model for the SN 1983 in NGC 1448. The internal accuracy and unambiguity of the model parameters has been studied by use of the equivalent widths and line strengths of the Balmer lines.

2. SOLUTION OF THE RADIATIVE TRANSFER PROBLEM

For the solution of the radiative transfer problem of the balmer lines we use a 10-level hydrogen atom. The population numbers N_i are given by the statistical equations (cf. Mihalas, 1978)

$$\frac{dN_i}{dt} = \Sigma N_j P_{ji} - N_i \Sigma P_{ij} = 0 \qquad (1)$$

where P_{ij} includes all radiative and collision processes, e.g. all bound-bound and bound-free transitions. The dependence of the radiative rates on the radiation field is given by the scattering integral. Here we assume complete redistribution, therefore the scattering integral is given by

$$\bar{J} = \int dx \int d\mu\, \phi(x) \quad I(x,\mu) \qquad (2)$$

with

x = frequency in Doppler units $(\nu_0 v_{th}/c)$

$\mu = \cos\theta$ (direction cosine)

ϕ = line profile function

I = specific intensity

The line source functions are given by

$$S_{ij} = \frac{2h\nu^3}{c^2} \left(N_i g_j / N_j g_i - 1\right)^{-1} \qquad (3)$$

and the line opacities can be written

$$\varkappa_{ij} = \frac{\pi e^2}{mc} f_{ij} \frac{\phi_\nu}{\Delta\nu_D} (N_i - N_j g_i/g_j) \qquad (4)$$

The comoving frame equation of transfer is given by equation (5) (cf. Mihalas, 1978)

$$\pm \frac{\partial I^{\pm}}{\partial z} - Q \frac{\partial I^{\pm}}{\partial x} = \varkappa_{ij} (S_{ij} - I^{\pm}) \qquad (5)$$

with

$$Q(p,z) = \mu^2 \frac{dv}{dr} + \frac{1-\mu^2}{r} v(r) \qquad (6)$$

Equation (5) is the radiative transfer equation along a ray through the atmosphere. The ray is specified by the impact parameter p.

Equation (5) has been solved by a numerical code which is similar to the method developed by Mihalas, Kunasz and Hummer (1975). The solution of the statistical equations and the transfer equation has been done by a J-N iteration scheme, which has been already used by Castor and van Blerkom (1970) for Wolf-Rayet stars. In this iteration scheme we start with an approximate solution which is given by the Sobolev solution (Sobolev, 1958, 1960; Castor, 1970) . From the Sobolev solution we get an approximate radiation field. With this radiation field we solve the statistical equations for the population numbers N_i . Line source functions and opacities can now be calculated by equations (3) and (4). With known line source functions and opacities, an improved radiation field can be obtained from equation (5).

The high expansion velocities in supernova atmospheres are responsible for small scattering zones. Therefore the radiation field is mainly controlled by local quantities. In this case the Sobolev solution is a very precise approximation in the atmosphere, but not at the boundaries. From equation (7) and (8), after convergence of the J-N iteration, we obtain the formal solution of the radiative transfer equation, e.g. the emergent flux:

$$I(x,p) = I_o \exp(-\tau_{MAX}) + \int_0^{\tau_{MAX}} S(r) \exp(-\tau) \, d\tau \qquad (7)$$

$$F(x) = \frac{2}{R^2} \int_0^R I(x,p) \, p \, dp \qquad (8)$$

with

$x = \bar{x} - \mu \, v(r)$

$\bar{x}$ = CMF frequency

x = observers frame frequency

The structure of the expanding SN atmosphere has been described by a core halo model, e.g. we did not calculate the transfer in the continuum. The inner

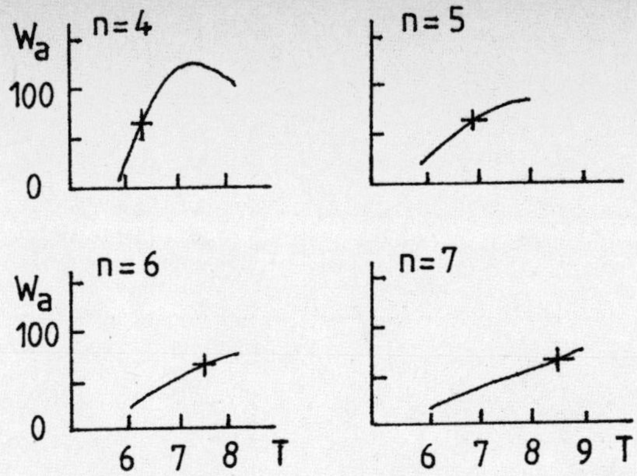

Fig. 1 The equivalent width of the absorption of H_α is plotted versus temperature. The equivalent width is measured in Å and the temperature is measured in units of 1000 K. Each diagram is labeled by the density exponent.

boundary of the atmosphere is given by the photosphere R_*. The velocity is as usually assumed to be proportional to the radius $v \sim r$. The density is described by a power law $\rho \sim r^{-n}$, and the temperature T is assumed to be constant throughout the atmosphere. The free parameters are the temperature T, the velocity at the photosphere v_o, the photospheric radius R_*, and the density exponent n. The density ρ_o at the photosphere is not a free parameter. ρ_o is determined by the condition $\tau \sim 1$ in the Balmer continuum (core-halo condition).

3. RESULTS

All observed Balmer line profiles can be described by 4 quantities: the equivalent widths of the emission and absorption, the strength of the emission, and the depth of the absorption. The same parameters can be determined from the theoretical calculations. For the determination of the best model we start our calculations with 3 fixed model parameters and plot the theoretical line parameters versus T as shown in Fig.1. From this diagram we find with the observed line parameters the temperature at which the model will match the observations. In the next step we vary the density exponent n and determine again the temperature. As a result of this procedure we get a set of 2 fixed and 2 varied model parameters (Fig. 1). Within this set the models match the observations. Now we can plot the temperature versus the density exponent n and get Fig.2. In

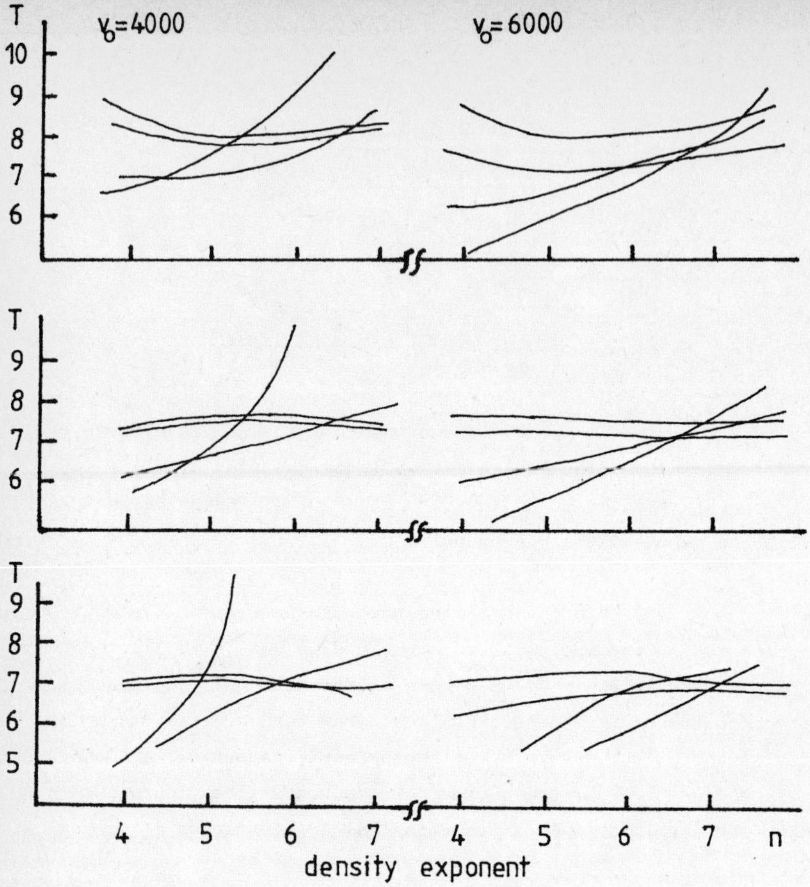

Fig. 2 Lines of constant equivalent width/absorption depth are plotted in a temperature-density exponent diagram. The temperature is measured in units of 1000 K. The left panel is calculated for a model with v_o =4000 km/s and the right panel corresponds to a model with v_o =6000 km/s.

this diagram lines of constant equivalent widths and line strengths are shown. Along these lines the observed line parameters are represented by the models. For each Balmer line we get 4 curves. The region of intersection of these curves defines the best values of temperature and density exponent for the fixed values of velocity and photospheric radius. The area of the intersection region depend on the fixed parameters. Therefore the fit procedure must be repeated for different values of the expansion velocity and the photospheric radius to minimize the area of intersection. In the ideal case the area will be reduced to

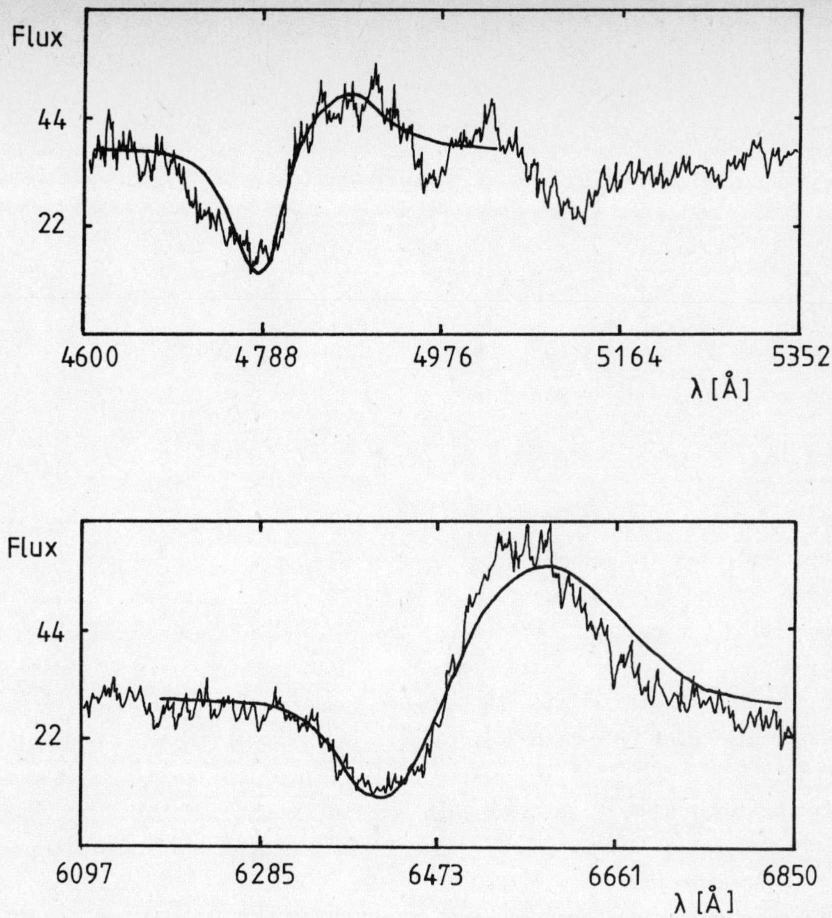

Fig. 3 Theoretical line profiles for the best model are compared with the observations of SN 1983 in NGC 1448. In the upper panel the H_β line and in the lower panel the H_α line are shown. Fluxes are given in arbitrary units.

a point, but this will never happen in real models. From the extension of the area of intersection we get the information on the internal accuracy of the model parameters.

In Fig.3 the H_α and H_β lines of the final model are compared with the observations. We find from the fit diagrams that nearly all model parameters can be determined with a very small internal error, but the photospheric radius R_* cannot be determined from the Balmer lines. Therefore the final model can be found only with additional informations. The age of the supernova, e.g. the time since explosion must be consistent with the velocity of the photosphere and the photospheric radius. The observations have been obtained 22 days after discovery of the SN. Therefore the age of the SN is $>$ 22 days and we find for our best model atmosphere the following structure:

$$
\begin{array}{lcll}
R & = & (1.0 \pm 0.5)10^{15} & \text{cm} \\
v_o & = & (5.5 \pm 0.2)10^{8} & \text{cm s}^{-1} \\
T & = & (7.25 \pm 0.4)10^{3} & \text{K} \\
n & = & 5.75 \pm 0.5 & \\
\text{age} & \gtrsim & 22 & \text{days} \\
\rho_o & = & 7.2\ 10^{9} & \text{cm}^{-3} \\
n_e & = & 7.08\ 10^{9} & \text{cm}^{-3}
\end{array}
$$

4. CONCLUSIONS

Model parameters of SN atmospheres obtained from spectroscopic fine analysis can be determined with an internal accuracy of about 10 percent, if the radius of the photosphere is known. The photospheric radius cannot be determined independently with an accuracy better than a factor of 2. For a distance determination, the velocity at the photosphere, the age of the SN or the photospheric radius must be determined. The velocity at the photosphere can be determined very accurately, but the age of the SN is somewhat uncertain, because the time between explosion and maximum light may be inaccurate by some days. Therefore the product of velocity and age of the SN may be accurate to within a range of 20 to 30 percent. On the other hand the photospheric radius can be determined from absolute flux measurements of the SNe. Unfortunately SNe are not black bodies, therefore the radius determination is again uncertain by a factor which might be larger than 2. Improvements in the distance determination using the Baade-Wesselink method can be made, if one calculates synthetic continuum models to determine absolute fluxes correctly.

5. ACKNOWLEDGEMENT

I am very indebted to O.-G. Richter for a copy of his La Silla observations.

REFERENCES

Axelrod,T.S.:1980, Ph.D. Thesis, University of California, Santa Cruz

Axelrod,T.S.:1981, in Proceedings of the NATO Advanced Study Institute on Supernovae, Cambridge England

Baade,W.:1926, Astr. Nachr. 228, 359

Branch,D.,Falk,S.W.,McCall,M.L.,Rybski,P.,Uomoto,A.K.,Wills,B.J.:1981, Astrophys. J. 244, 780

Branch,D.,Buta,R.,Falk,S.W.,McCall,M.L.,Sutherland,P.G.,Uomto,A., Wheeler,J.C.,Wills,B.J.:1982, Astrophys. J. 252, L61

Branch,D.,Lacy,C.H.,McCall,M.L.,Sutherland,P.G.,Uomoto,A., Wheeler,J.C.,Wills,B.J.:1983, Astrophys. J., 270, 123

Castor,J.I.:1970, Monthly Not.Roy.Astr.Soc. 149, 111

Fransson,C.:1984, Astron. Astrophys. 111, 140

Hempe,K.:1981, Astronom. Astrophys. 98, 19

Hempe,K:1983, Mitt. A.G. 60, 107

Mihalas,D.,Kunasz,P.B.,Hummer,D.G.:1975, Astrophys. J. 202, 465

Mihalas,D.:1978, Stellar Atmospheres (San Francisco:Freeman) p. 504

Sobolev,V.V.:1958, in Theoretical Astrophysics (New York:Pergamon Press)

Sobolev,V.V.:1960, in Moving Envelopes of Stars, Harvard University Press

Wesselink,A.J.:1949, B.A.N. 10,91

PHYSICAL MODELS FOR TYPE I SUPERNOVAE AND THE DISTANCE SCALE

J. Craig Wheeler
University of Texas at Austin

Peter G. Sutherland
McMaster University

Abstract

The degenerate carbon deflagration models for Type I supernovae are consistent with a variety of observations of light curves and spectra. Progenitor systems consisting of carbon/oxygen white dwarfs with normal companions or of binary white dwarfs are discussed and limits are established on the amount of ^{56}Ni which can be ejected in models which match the observations. The luminosity at maximum light in these models is proportional to the amount of nickel ejected, and hence limits can be set on the distance scale and H_0. If the carbon deflagration model is correct, then $0.4 < H_0/(100\ km/s/Mpc) < 0.7$.

1. Introduction

A number of recent studies (Sutherland and Wheeler 1984, Müller and Arnett, 1982, Nomoto, Thielemann, and Yokoi 1984, Woosley, Axelrod, and Weaver 1984, and references therein) have shown that the model for Type I supernovae based on the propagation of a carbon burning deflagration wave in a white dwarf is in good accord with a wide variety of observations. This model naturally produces about 1 $M_\odot$ of ^{56}Ni in an inner incinerated core and a few tenth's of a solar mass of C, O, and their burning products in an outer partially burned or unburned mantle.

The Ni undergoes radioactive decay by means of positrons and gamma rays with a half life of 6.1 day to ^{56}Co which in turn decays with a 77 day half life to ^{56}Fe. The decay of the Ni and Co gives a good account of the light curves (although there are still uncertainties in the treatment at intermediate and late times as the ejecta become optically thin) and the predicted physical conditions and abundances of Co and Fe give theoretical spectra in the late nebular phase which agree well with the observations (Axelrod 1980 and Woosley, Axelrod, and Weaver 1984). The incompletely burned mantle is a crucial feature of the model because it reproduces the spectra near maximum light very well (Branch et al. 1982, 1983 and contributions by Branch and Harkness in this volume).

In Section 2 the problems of explaining Type I supernovae as accretion of hydrogen onto carbon/oxygen dwarfs are reviewed. The scenario involving the evolution of two white dwarfs in close orbit is outlined and some dynamical calculations pertaining to this picture are presented.

Sutherland and Wheeler (1984) have established that to reproduce the light curves and particularly the kinematics of Type I events as deduced from spectral synthesis analyses, a minimum of 0.8 $M_\odot$ of C/O must be burned in the context of the carbon deflagration model. If less nuclear energy is released only a miniscule amount of matter will be moving at velocities of 10,000 to 12,000 km/s which are deduced for the matter at the photosphere at maximum light. In addition, the rise time to maximum light gets uncomfortably longer than 15 days.

In Section 3 arguments from a recent paper by Arnett, Branch and Wheeler (1985) are reviewed by which the minimum burning allowed in the carbon deflagration model can be used to set limits on H_0 which are independent of the properties of the atmosphere. These limits depend only on the apparent observed and predicted bolometric luminosities, not, for instance, on any knowledge or statement about the effective temperature.

2. Binary White Dwarf Model of Type I Supernovae

The fact that the carbon deflagration model is in accord with a variety of observations is reasonably well established. One of the remaining problems is to understand the stellar evolution which leads to this endpoint.

The simplest hypothesis is that one star of a binary pair forms a carbon/oxygen white dwarf and then the second star evolves and transfers mass to the dwarf raising its mass to the ignition point near the Chandrasekhar limit. Perhaps surprisingly, this basic idea has many problems in practice as pointed out by a number of authors (Fujimoto and Taam 1982, Sutherland and Wheeler 1984, Iben and Tutukov 1984, McDonald 1984).

Different problems arise in different regimes of mass transfer. They will be outlined here in only a brief schematic fashion. These arguments apply specifically to the transfer of hydrogen from a main sequence or giant star.

If mass is transferred to a C/O dwarf at a rate less than about 10^{-9} $M_\odot$/yr (the actual limit depending on the initial dwarf mass) a layer of degenerate hydrogen collects and finally ignites explosively and the dwarf undergoes a nova explosion and ejects the accumulated mass. Some small amount may be left, but the observations of excess CNO elements in nova ejecta (Ferland and Shields 1978) strongly suggest that some of the dwarf itself is ejected so that the mass of the dwarf may actually decrease.

If the mass transfer rate is in the range 10^{-9} to 5 x 10^{-8} $M_\odot$/yr, hydrogen burns quasistatically but a thick layer of degenerate helium accumulates. The helium is very volatile and when it ignites a detonation wave is sent out through the helium and into the C/O core incinerating the whole dwarf (Woosley, Weaver and Taam 1980, Nomoto 1982). This process can not lead to a classical Type I supernova because the resulting velocities are too high and there are virtually no intermediate mass elements left in the ejecta.

If the accretion rate is between 5 x 10^{-8} and 10^{-6} $M_\odot$/yr both H and He burn quasistatically, but the resulting luminosities are huge, in excess of 2000 $L_\odot$. To account for the rate at which Type I supernovae are thought to explode in the Galaxy, there must be of order a hundred such hydrogen burning white dwarfs within one kiloparsec of the sun, and no such systems are identified.

If the accretion rate is greater than 10^{-6} $M_\odot$/yr the transferred matter accumulates faster than it can be assimilated by a nuclear burning shell. Either the envelope will be ejected by radiation pressure so the core will not grow at all, or the envelope will linger and contaminate the resulting explosion with hydrogen, which is not observed in Type I supernovae.

Upon very close inspection, some of these problems may be overcome, but at the present time there is certainly no obvious way to make a Type I supernova based upon the accretion of hydrogen from a companion star onto a C/O white dwarf.

A possible resolution to this dilemma has been recently proposed by Iben and Tutukov (1984) and by Webbink (1984). These authors invoke the formation and evolution of close pairs of white dwarfs in binary systems to trigger the appropriate explosion. They describe a picture in which the originally more massive star forms a white dwarf. The second star evolves to form its own white dwarf core, but its red giant envelope swallows the first dwarf. A common envelope phase of evolution ensues in which the two dwarfs spiral together until their binding energy exceeds that of the envelope, and the envelope is expelled. The two dwarfs then spiral together under the influence of gravitational radiation until the less massive, which has the larger radius, fills its Roche lobe. That process can be catastrophic since as the dwarf loses mass, its radius swells forcing it even further beyond the Roche surface. The net result will be to disrupt the smaller dwarf and deposit its matter suddenly onto the larger dwarf.

Exploratory calculations of this very complex process are just now under way (Nomoto, Woosley, private communication), but it seems clear that the matter from the disrupted dwarf can not settle onto the larger dwarf without being heated by gravitational compression and igniting a burning shell. If the two dwarfs were initially of the same composition, burning the disrupted one in a shell will result in a composition inversion of a sort never before contemplated in stellar evolution. The result of the binary evolution of two helium dwarfs could be the formation of a layer of carbon and oxygen on top of a helium core. The evolution of a small mass helium dwarf and a larger mass C/O dwarf will just give a larger C/O dwarf, but the binary evolution of two C/O dwarfs will give a layer of O, Ne, and Mg on top of a C/O core as the carbon is burned in the shell.

With Li Zong Wei, we have done a series of calculations which attempt to set constraints on the configurations which may emerge from binary dwarf evolution.

The point is to determine which of various possible outcomes can serve as viable models for Type I supernovae. The models presume that a burning shell will consume all the accreted matter and that the burning shell may also penetrate to an arbitrary depth into the core representing the originally more massive dwarf. The mass of this core is taken to be a free parameter in the study.

We have determined that the evolution of two helium dwarfs is unacceptable as a model for Type I supernovae if the burning shell does not penetrate significantly into the inner core and the configuration undergoes central helium ignition. The helium is so volatile that the overlying mantle of C/O is also ignited. As is the case for other models based on helium detonation (Wheeler 1982) the resulting velocities are too large, and there is no mantle of unburned matter.

We have also calculated a series of models with C/O cores and O/Mg mantles. These models have a total mass of 1.4 $M_{\odot}$ and a deflagration is triggered in the center and allowed to propagate under the influence of the resulting Rayleigh-Taylor instability. We find that the deflagration penetrates the lower density, higher burning threshold mantle with great difficulty. For a C/O core of 0.19 $M_{\odot}$, only .47 $M_{\odot}$ burns, for a core of 0.47 $M_{\odot}$, only 0.57 $M_{\odot}$ burns, for a core of 0.73 $M_{\odot}$, only the core, 0.73 $M_{\odot}$ burns and for a core of .97 $M_{\odot}$, only 0.91 $M_{\odot}$, not quite all the core, burns. This means that if the burning shell penetrates too far into the central core, reducing the mass of C/O that can participate in the burning, the resulting explosion will be too mild to reproduce the velocities observed in Type I supernovae.

Figure 1 gives the final velocity profiles of the models with the C/O core masses as just specified. Of these, only the one with a core of 0.97 $M_{\odot}$ produces velocities in keeping with the observations of Type I supernovae. If the core is reduced to less than about 0.8 $M_{\odot}$ by the inward propagation of the burning shell, the explosion will be too feeble. If the burning shell penetrates to any appreciable depth in the inner core, the binary dwarf model will not be able to reproduce the requirements for a Type I supernova.

3. H_0 Without T_{eff}

The success of the carbon deflagration model in starting from a reasonably well understood physical model and reproducing self-consistently a variety of observational features encourages confidence in its use explore other issues, such as the distance scale. In particular, the model demands a minimum luminosity for Type I supernovae, and hence a lower limit to the distance and an upper limit to the Hubble constant.

As argued above, greater than 0.8 $M_{\odot}$ must be burned in order to provide the velocities observed in Type I supernova. Not all of the matter which is burned is turned into Ni because of the tendency for partial burning as the deflagration

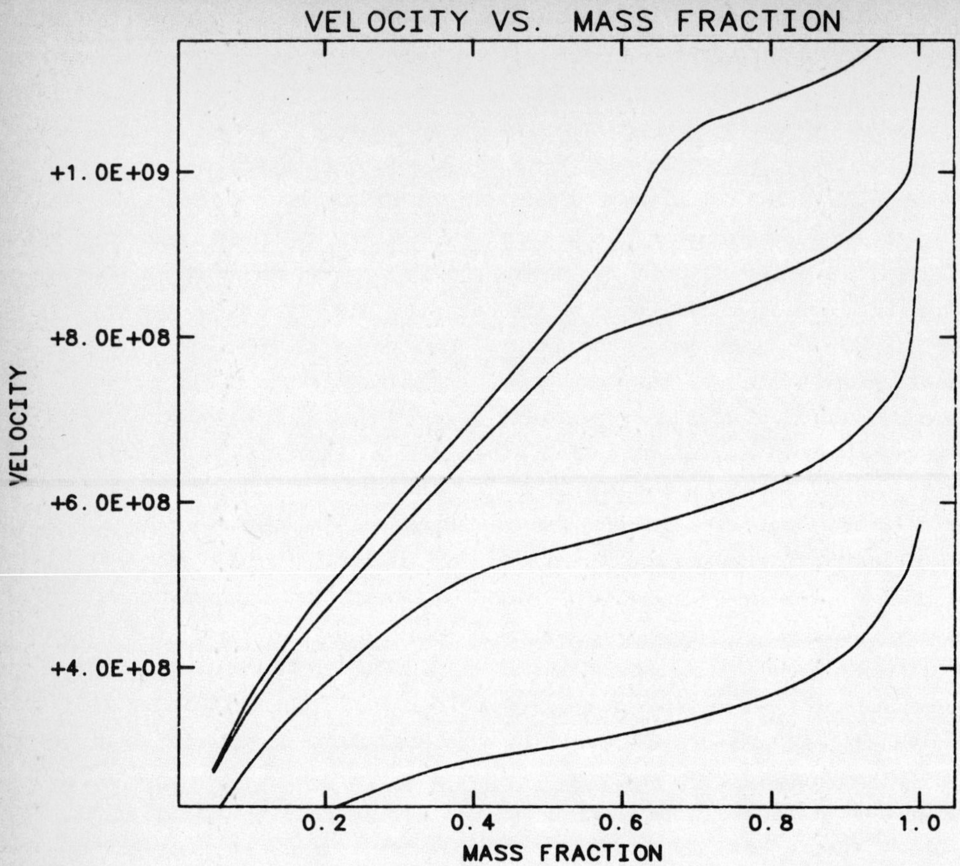

Figure 1. - The final velocity profile (in cm/s versus mass fraction) is given for a series of models predicated on the assumed outcome of the evolution of binary carbon/oxygen white dwarfs (see text). These models have a total mass of 1.4 M⊙. They are parametrized by the mass of a residual C/O core with the outer portions being a mantle of O and Mg. From top to bottom, the curves represent models with C/O cores of 0.97, 0.73, 0.47, and 0.19 M⊙. Only the model with the core of 0.97 M⊙, of which 0.91 M⊙ is burned in the explosion, produces velocities which can account for the observations of Type I supernovae.

fades. Müller and Arnett (1982), Nomoto, Thielemann, and Yokoi (1984) and Woosley, Axelrod, and Weaver (1984) find in detailed studies of the nucleosynthesis using both one and two dimensional numerical techniques that a fraction greater than 0.5 of the burned matter is incinerated all the way to ^{56}Ni.

Thus to give a model for a Type I supernova which is in accord with the observed velocities, the minimum amount of Ni which can be produced and ejected in the carbon deflagration model is 0.4 $M_\odot$. To give a comfortable amount of matter moving at velocities in excess of 10,000 km/s, and some matter at 20,000 km/s as implied by the width of features in the spectra, a model which burned about 1.2 $M_\odot$ of matter and produced of order 0.6 $M_\odot$ of Ni would be preferred. There is a natural upper limit to the amount of Ni which can be ejected, namely the Chandrasekhar limit 1.4 $M_\odot$. This limit is clearly extreme, since unburned matter, probably several tenths of a solar mass of it, is observed in the spectrum.

With numerical models the amount of Ni produced in a given explosion can be linked to the radiated luminosity at any epoch. Curiously, however, the time of maximum bolometric luminosity brings about a simplification in the sense that at that time the radiated luminosity is almost exactly equal, numerically, to the instantaneously deposited decay energy. This feature was first noted by Arnett (1982) in his study of analytic light curves, and is confirmed by numerical models (Sutherland and Wheeler 1984). This means that the absolute bolometric luminosity of the model at bolometric maximum can be determined simply from the rate of decay of the Ni present in a given model (the deposition function is very close to unity at maximum light).

Thanks to observations which span from the ultraviolet to the infrared, the apparent bolometric luminosity of Type I supernovae is known to reasonably high accuracy, as shown in Figure 2. Thus the empirical bolometric luminosity of Type I supernovae is known with no reference to the properties of the supernova atmosphere. In particular, this luminosity is independent of any statement about the effective temperature!.

Arnett, Branch and Wheeler (1985) have used these arguments based on deflagration in C/O white dwarfs to set limits on the Hubble constant (see also the paper by Branch in this volume). They invoke the empirical statement that the maximum B magnitude occurs about 15 days after the explosion, and hence that the maximum bolometric magnitude occurs at about 17 days after the explosion. The bolometric luminosity is then given by

$$L_{bol} = D\, \varepsilon_{Ni}(17d)\, M_{Ni} = 2.18 \times 10^{43}\ \mathrm{erg/s}/M_\odot\, M_{Ni}$$

where the deposition function, D, is nearly unity, the energy generation rate per gram from radioactive decay of Ni and Co is a known function of time, and M_{Ni}

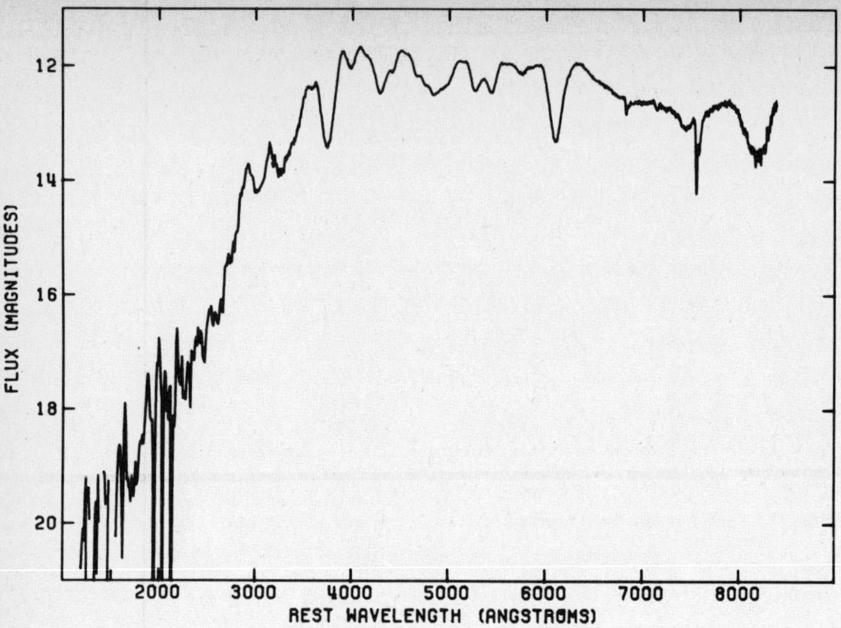

Figure 2. - The spectrum of Type I supernovae from the ultraviolet to the near infrared gives an empirical estimate of the apparent bolometric luminosity with no reference to the properties of the atmosphere or to an effective temperature. From Branch et al (1983) by permission of the Astrophysical Journal.

is the amount of ^{56}Ni produced in the explosion. This result is independent of the details of the composition, opacity, and deposition function.

With the limits on the mass of Ni ejected $0.4 < M_{Ni}/M_\odot < 1.4$ with a favored value of 0.6 $M_\odot$, the peak blue magnitude of Type I supernovae is predicted to fall in the range $-20.4 < M_B < -19.1$ with a favored value of -19.6 (see the paper by Branch in this volume). Using the redshift magnitude diagram for Type I supernovae in elliptical galaxies beyond the Virgo cluster (Sandage, Branch, this volume) one obtains $0.38 < h < 0.71$, with a favored value of 0.58.

If the progenitors of Type I supernovae are binary white dwarfs, the arguments are only slightly modified. With the outer mantle of O, Ne, and Mg, the energy released per gram is less. To get the same velocities, more matter would have to be burned. This would give more Ni, a higher luminosity, and hence a smaller Hubble constant. The models, on the other hand, show that very little of the heavier element mantle burns. Thus the only new constraint from this evolutionary picture is that the upper limit to the Ni ejection is reduced to approximately 1 $M_\odot$ and so the limits on the Hubble constant are more reasonably $0.45 < h < 0.71$. This argument is based on the assumption that the initial core mass is close to 1 $M_\odot$.

One must again emphasize that while these constraints on the distance scale do not depend sensitively on the details of the model, they do depend critically on the fundamental assumption of the model, namely that the energy of the explosion is directly related to the amount of Ni produced by the thermonuclear burning of carbon and oxygen. There is still a possibility that some as yet unexplored configuration based on helium detonation can be made to account for the observations. The greater energy liberated in helium burning would allow less Ni for a given kinetic energy even in the context of a thermonuclear explosion model. Although a model based on core collapse seems unlikely to reproduce all the observational features as successfully as the deflagration model, such a model would have a different source of kinetic energy, and could, in principle, eject less Ni.

Aknowledgements

We are grateful to David Arnett and David Branch for sharing their insights into the impact of supernova models on the distance scale, Ken'ichi Nomoto for discussions of presupernova evolution, and Doug Swartz for helping with the calculations. This research was supported in part by NSF Grant 8201210.

References

Arnett, W. D. 1982, Ap. J., 253, 785.

Arnett, W. D., Branch, D. and Wheeler, J. C. 1985, Nature, submitted.

Axelrod, T. S. 1980, Ph.D. thesis, University of California, Santa Cruz.

Branch, D., Buta, R., Falk, S.W., McCall, M.L., Sutherland, P.G., Uomoto, A., Wheeler, J.C., and Wills, B.J. 1982, Ap. J. Letters, 252, L61.

Branch, D., Lacy, C. H., McCall, M. L., Sutherland, P. G., Uomoto, A., Wheeler, J. C., Wills, B. J. 1983, Ap. J., 270, 123.

Ferland, G. J., and Shields, G. A. 1978, Ap. J., 226, 172.

Fujimoto, M. Y. and Taam, R. E. 1982, Ap. J., 260, 249.

Iben, I., Jr., and Tutukov, A. V. 1984, Ap. J. Suppl., 54, 335.

McDonald, J. 1984, in press.

Müller, E. and Arnett, W. D. 1982, Ap. J. Letters, 261, L109.

Nomoto, K. 1982, Ap. J., 257, 780.

Nomoto, K., Thielemann, F.-K., and Yokoi, K. 1984, in preparation.

Webbink, R. F. 1984, Ap. J. Lett., 277, 355.

Wheeler, J.C. 1982, in Supernovae: A Survey of Current Research, eds. M.J. Rees and R. J. Stoneham (Dordrecht: Reidel), 167.

Woosley, S.E., Axelrod, T.S., and Weaver, T.A. 1984, in Proceedings of the Erice Workshop on Stellar Nucleosynthesis, ed. C. Chiosi and A. Renzini, in press.

Woosley, S. E., Weaver, T. A., and Taam, R. E. 1980, in Type I Supernovae, ed. J. C. Wheeler (Austin: University of Texas), 96.

A Preliminary Discussion of Bose-Einstein Diffusion in Supernovae

Albert J. Fu and W. David Arnett
The University of Chicago
Department of Astronomy and Astrophysics

Introduction

In this report, a new method for the treatment of radiation dynamics in supernovae is discussed. The radiation field is characterized by a Bose-Einstein distribution with nonzero chemical potential. Radiative transport is treated by a diffusion technique. A bootstrap procedure is invoked to argue that the physical conditions thought to prevail during the explosion favor a Bose-Einstein radiation field for a significant portion of the evolution. A primary consequence of the theory may be that supernovae are dimmer than they are now thought to be and that distances are therefore smaller; values of H_0 inferred from supernovae would then be larger. An evaluation of H_0 based on the distance to NGC 4699 found from an independent technique is found indeed to be larger, by a factor of 2, than that predicted by an appropriate model for a blackbody supernova.

Physical Basis

Supernova explosions are highly complex, rapidly evolving events for which existing models have been shown to be useful in providing a clear picture of the hydrodynamic phenomena, accounting for the gross energetics, and fitting the shapes and features of light curves (Grassberg, *et al.* 1971; Chevalier, 1976; Falk and Arnett, 1977; Arnett, 1980 and 1982; Schurmann, 1983). With regard to the radiation dynamics, most of these models employ equilibrium diffusion techniques in the optically thick layers, 'equilibrium' meaning *Planck* equilibrium, and some match this to a transport scheme at small optical depths. But there are many problems related to a widespread application of Planck radiation to a supernova--

1) Although the typical absorption mean free path of a photon may be much less than the size of the object, i.e., $\tau_a >> 1$, where τ_a is the absorption optical depth of the object for a photon of energy $h\nu = kT_\gamma$, it is entirely possible that electron scattering will dominate the photon/matter interactions, i.e., that $\tau_{es} > \tau_a(h\nu = kT_\gamma)$. Furthermore, when $\tau_{es} >> \tau_a$, incoherent effects via multiple scattering are important. Felten and

Rees(1972) and Wagoner(1981) showed that coherent electron scattering will dilute an otherwise blackbody emergent spectrum while Kompaneets(1957) first showed that Compton scattering will thermalize(or rethermalize) a photon gas (if interactions are limited to electron scattering) to a Bose-Einstein distribution. What type of photon equilibrium prevails in the diffusive layers then depends crucially on the nature of the dominant opacity. It should be noted that because the chemical potential for a classical Bose gas is negative, a nonzero value of μ represents a ***dilution*** of Planck radiation at the same T_r

2) The establishment and maintenance of a photon equilibrium is largely a matter of kinetics in an object like the supernova, for which all quantities tend to vary on a rapid fluid flow timescale. Quick changes in density and temperature of the matter severely affect opacities, mean free paths, and photon creation and destruction rates. Even if we assume that a Planck equilibrium obtains for the initial configuration (which is static and for which densities are relatively high), the drastic decline of the absorption opacity as the expansion occurs renders the preservation of thermodynamic equilibrium of the radiation doubtful (barring the effects of 'expansion' opacity not considered here--see Karp, *et al.* 1977). Electron scattering may then produce a Bose-Einstein equilibrium if the relevant relaxation timescale can compete successfully against the electron scattering diffusion timescale.

3) Viewed as a problem in radiation hydrodynamics, the supernova problem amounts to solving the comoving frame equations of radiative transfer formulated by Castor(1972) for spherically symmetric flows. As a set of nonlinear coupled differential equations, any justifiable simplifications made at the outset are clearly desirable. Such a simplification would be the assumption of a universal functional form of the photon energy distribution. If we adopt a bootstrap procedure to solve the equations, that is, a process of conceptual iteration in which Bose-Einstein diffusion represents the first modification of present treatments, then a global description of the radiation as Bose-Einstein is desirable because it is a two-parameter fit instead of a one-parameter fit. In the transport regime, it is important to have independent parametrization of photon number and energy since these quantities in general vary independently, and a Bose-Einstein radiation field is certainly sufficient as a functional form. While diffusion may be an incomplete general representation of radiative transport, this two-parameter fit may actually be a good approximation. The Planck energy density must be chosen to make the number density come out right, and is inadequate. It is in this general sense that a BE equilibrium approximation seems to transcend its pure microphysical significance, and in a transport scheme takes on the character of an interpolation formula. The BE function also should be the ***physically appropriate*** two-parameter function. Since Bose statistics is a generalization of Planck statistics, given that we formulate the interaction terms in the dynamical equations in a reasonable way, a BE photon field is a self-consistent probe of the character of the

equilibrium. In particular, the chemical potential should become small in size whenever the physical conditions (generated by the model) warrant a Planck equilibrium.

Examination of Physical Conditions

One of the simplest and most straightforward things we can do is to adopt a specific model for a SN explosion, look at a set of numerical calculations based on the model in order to find out from the calculated evolution of temperatures, densities, and opacities which processes are important, and to use these findings to demonstrate the need for improvement. The model to be considered is the shock model of Falk and Arnett(1977) for a Type II explosion, in which we start with a 10^{51} erg piston at the bottom of an extended red supergiant envelope of $\sim$10 $M_\odot$. The movement of the shock through the envelope causes a nearly homologous expansion with a scale velocity approaching $\sim$0.10c. Radiative transport is given a diffusion treatment with a one parameter description of the radiation field, namely the temperature. Matter is assumed to be at all times in LTE. In order to concentrate on the radiative transfer we look at temperatures, densities, and opacities. Rosseland opacity curves for various densities in the supernova evolution are sketched in a crude fashion below and on them are drawn the (ρ, T) loci of the envelope at various times:

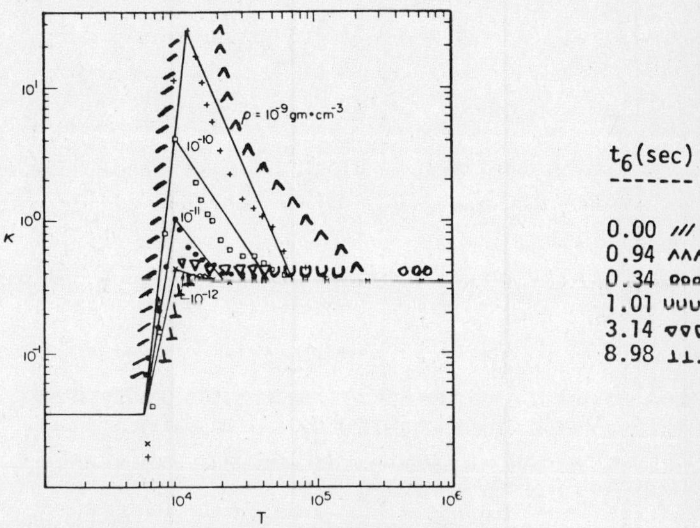

Temperature dependence of the opacity. The variation of total opacity κ (in $cm^2\ g^{-1}$; Cox and Stewart 1965, Cameron Mixture I) with temperature is shown for typical gas densities. The solid lines represent the chosen fits to the detailed opacities; actual values are plotted as separate points for each density value. The choice of the artificial ledge $\kappa = 0.1\ \kappa_{es}$ for $T \lesssim 6000$ K is arbitrary and was made for calculational convenience (see text for discussion of this point). The value $\kappa = \kappa_{es}$ obtains for high values of the gas temperature. For the purposes of these calculations, ionization and recombination effects are not explicitly treated, so that these effects enter the transfer calculations only through the corresponding opacity variation with temperature.

Rosseland mean opacities and SN loci. Opacities are from Cox and Stewart(1970); sketch is from Falk and Arnett(1977).

1. The initial configuration has a mean density of $\sim 10^{-8}$ g/cc and lies mostly on the absorption edge, $\bar{\kappa}_a > \kappa_{es}$.
2. The immediate effect of the shock is to heat the inner layers (as it passes) throwing them to the right in temperature coordinate by more than an order of magnitude, while rarifying these layers as it sweeps material out and accelerates it. The innermost layers are hot enough to lie on the Thomson portion of the opacity curve.
3. As the shock emerges, the configuration is left more or less isothermal at a temperature $\sim$ 20-30 times the initial mean temperature and the entire supernova lies on the Thomson plateau.
4. From this point on, the temperature drops off with a profile preserved by homology. But it is the density which dominates the opacity effects--because of the drop off in density, the diminishing of the absorption peak and the movement of the absorption incline will keep the SN neatly on the Thomson plateau all the way down to the recombination epoch.

This simple illustration shows that electron scattering is the dominant opacity (in terms of the transport mean free path) during the postshock evolution of the object, and becomes significant during the shock propagation phase.

Kompaneets(1957) showed that a photon gas inside a perfectly reflecting enclosure containing electrons would Comptonize to a BE distribution with a characteristic relaxation time of $t_{relax} = \frac{mc^2}{kT}\frac{\lambda_{es}}{c}$. Since a SN is neither an environment in which electron scattering is the only kind of interaction that occurs nor a perfectly reflecting enclosure we have to look at the criteria which indicate how well these conditions are simulated. We can state these most simply as

i) $\tau_{es} = \rho\kappa_{es}R >> 1$

ii) $t_{relax} << t_{diff}\,,\; t_{diff} \equiv \frac{R^2}{\lambda_{es}c}$

Because κ_{es} is independent of ρ and T and because the configuration is greatly extended, the first criterion will be satisfied until very late times. Using the fairly representative values of $\rho \approx 10^{-9}$ g/cc and $T \approx 10^5\,{}^\circ K$ for the immediate postshock adjustment epoch (Falk and Arnett 1977), we see that $t_{relax} = \frac{mc^2}{kT}\frac{1}{\rho\kappa_{es}c} \approx 10^4$ sec at this time in the evolution, while $t_{diff} \approx 1$ year, but the steadily dropping density will cause $t_{relax} \approx t_{diff}$ when $R \approx 3\times10^{15}$cm, or $t \approx \frac{(R-R_o)}{v_{sc}} \approx 3\times10^6$ sec corresponding to the postpeak decline on the theoretical light curve. After this, we can expect incoherent scattering to be no longer important as a thermalizing mechanism. We also note that by rearranging terms in ii) we can rewrite it as

$$\frac{kT}{mc^2}\tau_{es}^2 >> 1$$

or

$$Y_{NR} >> 1/4$$

namely that the nonrelativistic Compton Y parameter must be much greater than unity. Thus, incoherent scattering is an important energy redistribution mechanism for nonrelativistic electrons only through *multiple* interactions.

Frequency Dependent Character of the Radiation Field

We must consider that the supernova envelope is a finite medium and that absorptive and emissive mechanisms are, in general, frequency dependent. In particular, if we consider free-free opacity as the dominant absorption, the the strong ν^{-3} dependence will effectively make the SN a blackbody enclosure whose walls are high bandpass filters, so that low energy photons for which $\kappa_{ff} >> \kappa_{es}$ will tend to be maintained in a Planck equilibrium, whereas higher energy photons for which $\kappa_{ff} << \kappa_{es}$ will have their spectrum modified by incoherent scattering effects and then escape the medium via scattering. Indeed, in view of this *frequency dependent equilibrium*, we must make more stringent the previous criterion for establishment of a BE equilibrium:

$$t_{relax} << \frac{l(\nu)^2}{\lambda_{es}c}$$

where $l(\nu)$ is the destruction length of a photon of energy $h\nu$ and is easily found to be approximated by

$$l(\nu) \approx (\lambda_{es}\lambda_{ff})^{1/2}$$

whenever $\lambda_{es} << \lambda_{ff}$. Photons of decreasing energy will cease to satisfy this condition when

$$t_{relax} \approx \frac{l(\nu_{coh})^2}{\lambda_{es}c}$$

which defines the frequency ν_{coh} below which incoherent effects are not important (this being valid, of course, only as long as $l(\nu) \leq R$). Rearranging terms as before,

$$Y_{NR}(\nu_{coh}) >> 1$$

where $Y_{NR}(\nu) = \frac{4kT}{mc^2}[\tau_{es}(\nu)]^2$ and $\tau_{es}(\nu) = \rho\kappa_{es}l(\nu)$. Finally, scattering is not important at all below the frequency ν_o for which $l(\nu_o) = \lambda_{ff}$, or $\kappa_{ff}(\nu_o) = \kappa_{es}$, and absorption is not important at all above ν_t, where $l(\nu_t) = R$, or when

$$\frac{\lambda_{ff}(\nu_t)}{c} > \frac{R^2}{\lambda_{es}c}$$

that is, when the mean survival time for a photon exceeds the scattering diffusion time for the entire region. A discussion of these frequencies is given in Rybicki and Lightman(1979).

These considerations, of course, have important effects on the character of the *emergent* spectrum. Felten and Rees(1972) first showed that coherent electron scattering would modify a blackbody emergent spectrum, essentially by decreasing the effective volume in the outer regions in which thermally emitted photons would escape the object. Wagoner(1981), from a different point of view, obtained a similar result by employing the Eddington approximation to find the emergent flux and noting that the optical depth at which the spectrum forms actually *increases* because of scattering even though $l(\nu) < \lambda_{ff}(\nu)$. The reason for this is that while $l(\nu)$ is less than $\lambda_{ff}(\nu)$ by the factor $\sim(\lambda_{es}/\lambda_{ff})^{1/2}$, the transport mean free path decreases when scattering dominates by $\sim\lambda_{es}/\lambda_{ff}$, increasing τ from ~ 1 to $\sim(\lambda_{ff}/\lambda_{es})^{1/2}$. Both approaches give *dilutions* of the otherwise blackbody flux. Wagoner went further to show that the emergent flux is given in terms of two parameters which indicate the magnitude of the dilution. We again can use the fairly representative values of $\rho \approx 10^{-9}$ g/cc and $T \approx 10^5\,^{\circ}K$ in the early postshock adjustment phase to evaluate the parameters $x_o \equiv \frac{h\nu_o}{kT}$ and $x_{coh} \equiv \frac{h\nu_{coh}}{kT}$ and find that $x_o \approx 0.1$ whereas $x_{coh} \approx 50$, so that coherent scattering is significant, but that incoherent effects are negligible. However, as the density drops off steadily during the postshock phase x_{coh} can be expected to drop below unity. A universal two-parameter description of the radiation in terms of μ and T would then be desirable because not only would it be able to account for flux dilution at all times but it would go over to a physically clear limit as x_{coh} becomes small, provided, of course, that $x_{coh} < x_t$.

Dynamical Equations

Castor(1972) formulated a relevant set of moment equations for spherically symmetric radiating fluids correct to order $\frac{v}{c}$. In frequency integrated form, they are:

$$\frac{DE_o}{Dt} + 4\pi\rho\frac{\partial(r^2F_o)}{\partial M} - \frac{v}{r}(3P_o - E_o) - \frac{D\ln\rho}{Dt}(E_o + P_o) = \int_0^\infty [4\pi\eta_o(\nu_o) - c\rho\kappa_o(\nu_o)E_o(\nu_o)]\,d\nu_o$$

$$\frac{1}{c^2}\frac{DF_o}{Dt} + 4\pi r^2\rho\frac{\partial P_o}{\partial M} + \frac{3P_o - E_o}{r} - \frac{2}{c^2}\left(\frac{v}{r} + \frac{D\ln\rho}{Dt}\right)F_o = -\frac{1}{c}\int_0^\infty \rho\kappa_o(\nu_o)F_o(\nu_o)\,d\nu_o$$

where the subscript $_o$ means that all quantities are evaluated in the fluid frame. In addition to imposing the diffusion constraint, so that the radiation field is zeroth order isotropic and $3P_o \approx E_o$, we impose the constraint of *dynamic diffusion* (Mihalas 1984),

$$t_{fluid} \ll t_{diff}$$

or

$$\frac{l}{v} \ll \frac{l^2}{\lambda_\gamma c},\ \frac{\lambda_\gamma}{l} \ll \frac{v}{c}$$

in other words, all quantities evolve on a fluid flow timescale. Here, λ_γ is the transport mean free path and l is a length scale of the flow. A simple dimensional analysis of the remaining three terms on the left hand side of the radiation momentum equation indicates that these terms scale like $(\frac{v}{c})(\frac{\lambda_\gamma}{l})$: 1 : $(\frac{v}{c})(\frac{\lambda_\gamma}{l})$ relative to the term on the right hand side, suggesting that we ignore everything but the pressure gradient term, and we end up with the diffusion transport equation

$$\frac{\partial P}{\partial r} = -\frac{1}{c}\int_0^\infty \rho\kappa F d\nu$$

where the subscripts have been dropped for convenience. Note that in this limit all the velocity dependent coupling terms go away. We can also manipulate the radiation energy equation to read (Mihalas 1984)

$$\frac{D}{Dt}\left(\frac{E}{\rho}\right) + P\frac{D}{Dt}\left(\frac{1}{\rho}\right) = \frac{1}{\rho}\int_0^\infty (4\pi\eta - c\rho\kappa E)d\nu - \frac{1}{\rho r^2}\frac{\partial}{\partial r}(r^2 F)$$

which is the first law of thermodynamics for the radiation. The complete set of equations of radiation hydrodynamics in the diffusion regime on a fluid flow timescale for spherical geometry would then be

energy

$$\dot{E}_m + p_m\dot{V} = \text{absorptions} - \text{emissions} + \text{energy generation}$$

$$= \frac{1}{\rho}\int_0^\infty (c\rho\kappa E_\gamma^v - 4\pi\eta)d\nu + \epsilon \qquad \text{(matter)}$$

$$\dot{E}_\gamma + P_\gamma\dot{V} = \text{emissions} - \text{absorptions} - \text{flux out of specific volume}$$

$$= \frac{1}{\rho}\int_0^\infty (4\pi\eta - c\rho\kappa E_\gamma^v)d\nu - \frac{1}{\rho r^2}\frac{\partial}{\partial r}(r^2 F_\gamma) \qquad \text{(radiation)}$$

momentum

$$\dot{v} = -\frac{GM}{r^2} - \frac{\partial}{\partial r}(p_m + P_\gamma) \qquad \text{(matter)}$$

$$\frac{\partial P_\gamma}{\partial r} = -\frac{1}{c}\int_0^\infty \rho\kappa F_\gamma d\nu \qquad \text{(radiation)}$$

photon number

$$\frac{D}{Dt}\left(\frac{n_\gamma}{\rho}\right) = \frac{1}{\rho}\int_0^\infty \frac{1}{h\nu}(4\pi\eta - c\rho\kappa E_\gamma^v)d\nu - \frac{1}{\rho r^2}\frac{\partial}{\partial r}r^2\int_0^\infty \frac{F}{h\nu}d\nu$$

where now E is matter specific and E_γ^v is the volume energy density. These three conservation laws of matter energy, photon energy, and photon number are required to describe the general two-component (matter, radiation) fluid with a 3-parameter (T_m, T_γ, μ) thermodynamic prescription.

The expressions for the energy density and number density are given exactly by

$$E_\gamma^v = \frac{8\pi h}{c^3}\int_0^\infty \frac{\nu^3 d\nu}{e^{\frac{\epsilon-\mu}{kT_\gamma}}-1} = \frac{48\pi}{h^3c^3}e^{\frac{\mu}{kT_\gamma}}\sum_{j=0}^\infty \frac{e^{\frac{j\mu}{kT_\gamma}}}{(j+1)^4}(kT_\gamma)^4$$

and

$$n_\gamma^v = \frac{8\pi}{c^3}\int_0^\infty \frac{\nu^2 d\nu}{e^{\frac{\epsilon-\mu}{kT_\gamma}}-1} = \frac{48\pi}{h^3c^3}e^{\frac{\mu}{kT_\gamma}}\sum_{j=0}^{\infty}\frac{e^{\frac{j\mu}{kT_\gamma}}}{(j+1)^3}(kT_\gamma)^3$$

which, for computational purposes, are approximated at *all* values of μ by the Maxwell-Boltzmann limit. For instance, E_γ^v becomes

$$E_\gamma^v \approx \bar{a}e^{\frac{\mu}{kT_\gamma}}T_\gamma^4$$

$$\bar{a} \equiv \frac{48\pi k^4}{h^3c^3}$$

and similary for n_γ^v. This expression is exact in the nondegenerate limit, and incurs a maximum error of $\sim$8 % in the degenerate(Planck) limit, since $\frac{\bar{a}}{a} \approx 0.92$, where a is the black-body value.

A major source of difficulty stems from the fact that in a two-component fluid description we must explicitly evaluate the photon/matter interaction integrals over frequency. In view of the above discussion, an appropriate formulation might be

$$\int_0^\infty (4\pi\eta - c\rho\kappa E_\gamma^v)d\nu = N_o^2\rho^2T_e^{-1/2}[A_{emi}\int_{\nu_{coh}}^{\infty}\bar{g}_{ff}(\nu)e^{\frac{-h\nu}{kT_e}}d\nu - A_{abs}\int_{\nu_{coh}}^{\infty}\bar{g}_{ff}(\nu)\frac{1-e^{\frac{-h\nu}{kT_e}}}{e^{\frac{h\nu-\mu}{kT_\gamma}}-1}d\nu]$$

where A_{emi} and A_{abs} are numerical constants obtained from the bremsstrahlung emissivity and free-free absorption cross section, and LTE is assumed for the matter ($T_e = T_m$). Below the frequency ν_{coh} there are no net sources since $E_\gamma^v \approx \frac{4\pi B_\nu(T_e)}{c}$. Such a formulation would take into account the effect of electron scattering on the radiation field, particularly on the emergent spectrum. Perhaps more appropriate would be a formulation which delineated the effects of incoherent scattering,

$$\int_0^\infty (4\pi\eta - c\rho\kappa E_\gamma^v)d\nu = N_o^2\rho^2T_e^{-1/2}[A_{emi}\int_{\nu_{coh}}^{\infty}\bar{g}_{ff}(\nu)\frac{e^{\frac{-h\nu}{kT_e}}}{h\nu}kT_e d\nu - A_{abs}\int_{\nu_{coh}}^{\infty}\bar{g}_{ff}(\nu)(1-e^{\frac{-h\nu}{kT_e}})e^{\frac{\mu}{kT_\gamma}}e^{\frac{-h\nu}{kT_\gamma}}d\nu]$$

where the last emission term is a Comptonized bremsstrahlung spectrum, approximated by shifting all the bremsstrahlung photons to kT_e (Rybicki and Lightman 1979). In the event that $x_{coh} << 1$, the first term should be small enough to be neglected. Of course, ν_o and ν_{coh} would both be found self-consistently; in a numerical calculation, values of ρ and T_e at the nth timestep would be used to evaluate ν_o and ν_{coh} at the (n+1)th timestep.

Such integrals not only represent a formidable computational rpoblem by themselves but make the whole numerical structure of the equations cumbersome. This is essentially because the delineating frequencies will vary from zone to zone and in time. Further analysis of the interaction terms is therefore of the utmost priority in performing an initial set of calculations.

Finally, we can comment on two other points. First of all, in Type II models, we can set $\epsilon=0$ in the matter energy equation; in Type I's, we would include ϵ as a time dependent radioactive decay source term. Secondly, in the transport regime, when dynamic diffusion breaks down, we must include the $\partial F/\partial t$ term in the photon momentum equation (or introduce a flux limiter in the time-independent diffusion equation).

The Modified Eddington Approximation

We can now consider a simple scheme to determine distances from the emergent flux. For blackbody radiation, the Eddington approximation for the atmosphere is

$$T^4 = \frac{3}{4} T_{eff}^4 (\tau + 2/3)$$

with the effective temperature defined by $L = 4\pi R_{eff}^2 \sigma T_{eff}^4$. For the case of BE radiation, we can approximate the energy density as discussed before by the nondegenerate limit:

$$E_\gamma^v(\mu, T_\gamma) = \bar{a} e^{\mu/kT_\gamma} T_\gamma^4$$

and defining T_{eff} the same way

$$\bar{\sigma} T_{eff}^4 \equiv \frac{L}{4\pi R_{eff}^2}$$

with $\bar{\sigma} \equiv \frac{\bar{a}c}{4}$. We thus end up with a *Bose-Einstein Eddington approximation:*

$$e^{\mu/kT_\gamma} T_\gamma^4 = \frac{3}{4} T_{eff}^4 (\tau + 2/3)$$

What implications would this have? Consider the following method of determining distance:

i) From $\bar{\sigma} T_{eff}^4 = L/4\pi R_{eff}^2 = L/4\pi d^2 (d^2/R_{eff}^2) = F_{obs} d^2/R_{eff}^2$, we get

$$d^2 = \frac{\bar{\sigma} T_{eff}^4 R_{eff}^2}{F_{obs}} = \frac{\sigma(\bar{\sigma}/\sigma) T_{eff}^4 R_{eff}^2}{F_{obs}}$$

$$\approx \frac{0.92\sigma T_{eff}^4 R_{eff}^2}{F_{obs}} = \frac{0.92\sigma T_{eff}^4 e^{\frac{\mu}{kT_\gamma}} R_{eff}^2}{F_{obs}}$$

ii) Estimate R_{eff} from the observed photospheric velocity and the observed time since the start of the coasting phase,

$$R(t) = v_{sc}(t - t_o) + R_o \approx v_{sc}(t - t_o)$$

where $R(t) >> R_o$.

iii) Formulate a conversion procedure to translate a color index, such as B-V, to a color temperature, T_c; T_c is defined as a spectral shape temperature, and so is directly related to a spectral shape factor such as a color index. For a blackbody continuum, T_c should in principle be equal to T_{eff} (at $\tau = 2/3$); but in this case, $T_c = T_\gamma$

An important point here is that the dilution effect is lumped almost entirely into the chemical potential with T_γ parametrizing the spectral shape. A certain B-V should then translate into about the same T_c for both the blackbody and the BE radiation. The difference is that a certain value of T_c for BE implies a lower value of T_{eff} than for a blackbody, i.e., the emergent flux is smaller. The derived distance d is therefore smaller.

An Example: SN 1983k in NGC 4699

Optical observations of the Type II supernova 1983k in NGC 4699 were presented by Niemela, *et al.* (1984). The extended peak of the lilght curve, the spectra near maximum light, and the values of the photospheric velocities measured from absorption lines led the authors to consider model F of Falk and Arnett(1977) as a likely candidate to best fit the data, the model having a 7 M⊙ red supergiant with a 1.7 M⊙ circumstellar shell as a progenitor. The following figure shows a comparison of the theoretical light curve of model F with that obtained by the observers:

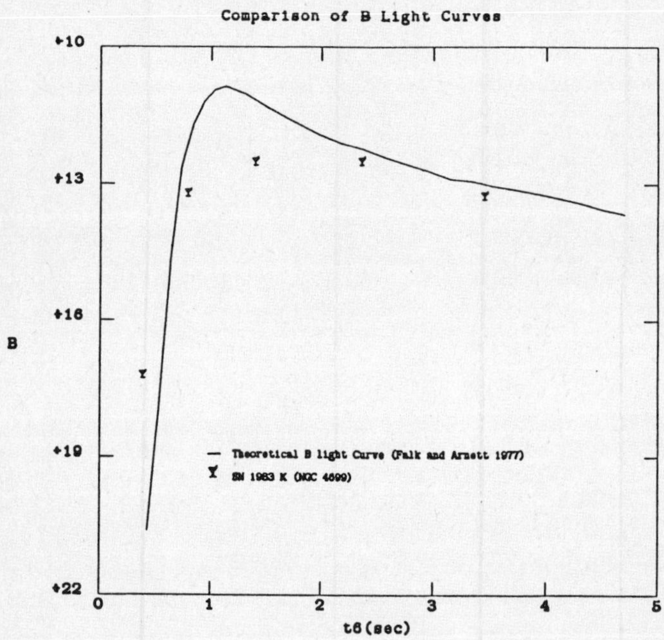

Comparison of theoretical and observational light curves.

Use was made of the distance modulus to NGC 4699 found by an independent method (de Vaucouleurs 1975) to convert the theoretical curve to apparent magnitudes. For the purposes of comparing B fluxes at various times of the evolution, we note that the matching of the light curves in time coordinate can be accomplished by lining up the early rising portion of the peaks of the curves, since these points are well defined landmarks for both curves. Because the observed peak is dimmer, a dilution is suggested, but the close matching of the postmaximum portions suggests that it may be a small effect. In addition, other models can possibly match the observed shape better, and it would probably be desirable to construct a detailed atmospheric model to generate UBV. Nevertheless, let us compare the B fluxes at maximum light, since these are the only points that can be compared without interpolating the tabulated data for the model light curve. We find that the discrepancy in B magnitude, hence in distance modulus, is $12.^m4$(obs) - $10.^m9$(theor) $\approx$ 1.5. The distance modulus predicted from the blackbody supernova is therefore $\sim$ 31.9, which amounts to a distance overestimate by roughly a factor of $10^{1.5/5} \approx 2$. This suggests that H_o would then be larger by the same factor. The above illustration cannot produce a serious value of H_o, or give reason for choosing one investigator's H_o over another's, but it does show the possible importance of the effect discussed; in this case, the de Vaucouleurs value of H_o would seem to be more consistent with the diluted flux model than smaller values of H_o.

Optically Thin Transport and the Importance of μ

The postshock evolution is *adiabatic* in the sense that only a small fraction of the initial supernova energy is lost in the first several months of evolution. Indeed, when the shock emerges, t_{fluid} is about 1 day, whereas $t_{diff} \approx 1$ year. Thus, if we view the explosion as a *hydrodynamic* event, and in terms of gross energetics, then the supernova, at least for several months, looks like a radiating fluid in which the photons and matter are locked together. Since the expansion is also homologous, such a slowly diffusing radiation field can in fact be parametrized by a single parameter, the temperature, which at all points decreases with time as the fluid is diluted by the expansion, but whose profile is preserved by homology. From this point of view the explosion is not unlike the expansion of the universe as a whole: if we assume that at early times thermodynamic equilibrium prevails, perhaps not an unreasonable assumption for even the shock propagation epoch in which densities are relatively high, then a Planck distribution also prevails. At later times, as the *absorption* opacity drops sharply, the Planck distribution *freezes out.* But the decoupling transition in cosmology is different from our situation in one important respect--the universe is homogeneous everywhere in radiation and matter, and so a single temperature applies everywhere; in our case, the SN occupies a finite region of space and is highly inhomogeneous, and thus a temperature gradient is required to parametrize the inhomogeneity. However, as a radiating object the SN must make departures from the above simple picture. These departures must be considered significant since it is the loss of radiative energy which enables us to observe SN's and to test theoretical models. From this point of view, a SN continually loses photons in significant amounts; as the photospheric radius recedes in mass coordinate as the density drops off (Falk and Arnett 1977), the SN eventually becomes transparent. Therefore, in the optically thin laters photons tend to stream out of the object and a greater proportion of the object becomes optically thin as time goes on. Because of the sparseness of the interactions there is no energy redistribution mechanism and the photon temperature tends to freeze out, while no new photons are produced. But because of spherical dilution effects, i.e., because photons are moving out in radius and continuously filling larger and larger volumes of spce, the number density must drop off with radius. The claim is that this can be accomodated by a BE gas with a chemical potential gradient, the magnitude of the chemical potential becoming larger in the direction of the freeze out. Note that in cosmology, it is not the homogeneity of the universe but the effect of *redshift* in the expansion which allows the temperature to account for both a spatial dilution and an energy redistribution.

Conclusions

It has been argued that Bose-Einstein diffusion is a conceptual improvement in the treatment of radiative transport in supernovae. First and foremost, it represents a two-parameter description of the radiation field, the two parameters being μ and T_γ, whose associated *integrals*, the total photon number and the total photon energy, can be budgeted independently. As the first iteration in a bootstrap process to achieve an internally consistent SN model and to develop general techniques of handling radiative transport at all optical depths, it seems clear from the above discussion that such a two-parameter fit is a step toward these goals. Secondly, the microphysics of the relevant photon/matter interactions in the SN strongly suggests that the appropriate two-parameter function is indeed Bose-Einstein, since it gives the correct physical limits of photon equilibrium whenever the appropriate conditions as dictated by the laws of statistical mechanics apply. A major difficulty is the explicit formulation of the interaction terms in the dynamical equations. These integrals over frequency must be made computationally tractable and require more knowledge of the energy spectrum of the radiation. The simplest approach would be to employ the *nondegenerate* BE energy distribution; however, it should be kept in mind that although the frequency integrated *moments* of a Maxwell-Boltzmann gas can approximate the degenerate limit with an 8 % error, the corresponding energy distribution is a poor approximation of the Rayleigh-Jeans portion of the blackbody spectrum. Nevertheless, such an approach may be preferable in a first set of calculations.

REFERENCES

Arnett, W.D. 1980, *Ap. J.,* **237**, 541.

______ 1982, *Ap. J.*, **253**, 785.

Castor, J.I. 1972, *Ap. J.,* **178**, 779.

Chevalier, R.A. 1976, *Ap. J.,* **207**, 872.

Cox, A.N., and Stewart, J.N. 1970, *Ap. J. Suppl.* (No. 174), **19**, 243.

de Vaucouleurs, G. 1975, in *Stars and Stellar Systems,* vol. **9**, ed. A. and M. Sandage and J. Kristian (Chicago: University of Chicago Press), p. 557.

Falk, S.W., and Arnett, W.D. 1977, *Ap. J. Suppl.,* **33**, 515.

Felten, J.E., and Rees, M.J. 1972, *Astron. and Astrophys.,* **17**, 226.

Grassberg, E.K., Imshennik, V.S., and Nadyozhin, D.K. 1971, *Ap. Space Sci.,* **10**, 28.

Karp, A.H., Lasher, G., Chan, K.I., and Salpeter, E.E. 1977, *Ap. J.*, **214**, 161.

Kompaneets, A.S. 1957, *Soviet Phys.---JETP,* **4**, 730.

Mihalas, D., and Mihalas, B.W. 1984, *Foundations of Radiation Hydrodynamics* (in press).

Niemela, V.S., Ruiz, M.T., and Phillips, M.M. 1984 (in press).

Schurmann, S.R. 1983, *Ap. J.,* **267**, 779.

Wagoner, R.V. 1981, *Ap. J.(Letters),* **250**, L65.

SUPERNOVAE UP CLOSE
(An After-Dinner Talk)

Brian G. Marsden
Harvard-Smithsonian Center for Astrophysics

I am very honored to have been invited to speak at this workshop on supernovae, but I rather wonder what I am doing here. "What," I hear you say, "does Marsden know about supernovae?". Well, the answer is: not very much. Oh, I remember learning, long ago in graduate school, even in an astronomy department that actually specialized in, of all things, celestial mechanics, that type I supernovae occurred among population II stars, and vice versa; and that the type I and type II supernovae had absolute magnitudes of -18 and -20, not necessarily respectively. But, as far as using supernovae as distance indicators is concerned, that's the extent of my knowledge. Actually, from some of the discussion this morning, it seems that things haven't changed all that much in the past 20 years.

Then I wondered if there were some particular reason for having the supernova meeting at this time. Was there perhaps a special anniversary, or something? Of course, as everybody knows, there was a stupendous supernova in our galaxy in 1572, but that will hardly do: we already had here the colloquium to celebrate the quatercentennial of Tycho Brahe's discovery of that object. (Was that really twelve years ago?) And as for the famous supernova in the Andromeda galaxy, I should be jumping the gun if I thought that this meeting were being held in celebration of that; we must wait another whole year for the centennial of the first supernova to have both its photograph and its spectrum taken. Then I realized why I had been asked to speak to you this evening: modern research on supernovae really began with Fritz Zwicky's discovery of what was, by one tenth of a magnitude, the brightest supernova to occur this century; and Fritz Zwicky happened to make that discovery just a few days after I was born! It was that event--the appearance of the brightest supernova, not of me--that allowed Rudolph Minkowski to make the first detailed spectroscopic observations of a supernova, and <u>that</u> in turn inspired Cecilia Payne-Gaposchkin and Fred Whipple to produce the first theoretical treatment of just what supernovae are all about.

Supernovae are things that happen to other galaxies. At least, they have all been happened to other galaxies starting with the one I just mentioned in Messier 31 ninety-nine years ago. That event of 1885 is the first entry in the catalogue of supernovae that now contains, I believe, about 574 entries. Other galaxies tend to have names like NGC 1234 or IC 321. When someone discovers a supernova, he may say, sometimes to his colleagues but usually only to himself, "I see a supernova in IC 812!"--and then he sends (or gets his director to send) us a telegram. The telegram would generally be a little more informative than his statement. The telegram might read something like: N4757 SUPERNOVA BARTEL 19501 40927 12503 11002 04190 10020 20015 18098 18158 27695 SHAPIRO (you're not writing this down, are you?). Then, with the speed of light, that string of words and numbers would come clattering in to the teletype machine in my office. If I'm not too busy, I might perhaps stroll over to the machine to have a look and say, possibly to my colleagues but usually only to myself, "Oh, look! The director of such-and-such observatory (I would probably know which observatory it was) is telling us that one of his flun-- staff members has found a supernova in NGC 4757. Big deal! I suppose they know what they're doing. I wonder if 4757 is a prime number. Well, perhaps we'd better relay the telegram to other observatories in case anyone else wants to look at it... though why anybody else would be interested in a nineteenth-magnitude supernova, I do not know!" We would probably look to see if NGC 4757 is in de Vaucouleurs' excellent catalogue of bright galaxies and check that the position de Vaucouleurs gives for the galaxy agrees with what Bartel and Shapiro coded into their telegram.

Actually, most galaxies are not in the de Vaucouleurs catalogue: the catalogue may be excellent, but it is not perfect. Most galaxies do not even have the convenient NGC or IC numbers. Some galaxies might be in obscure places like Vorontsov-Velyaminov's "Morphological Catalogue" and the listings by Zwicky et al. from the Palomar Sky Survey charts; or perhaps from the more recent ESO survey of the southern sky. In such cases the galaxy will probably be described as "anonymous". Most supernovae are in anonymous galaxies. If some of you are trying to verify whether 4757 is a prime number, let me give you a clue. The clue is: 3. No, I know you have figured out that 4757 is not divisible by 3 (or even 2 or 5), but 3 is nonetheless a clue. Pocket calculators are not allowed for this exercise; those of you who might be inclined to use them are kindly requested to bring them up and deposit them on the table here, right in front of me.

Supernovae are often found in the course of special photographic patrols, such as those at Abastumani, Asiago, Budapest and Cerro El Roble. Some of the most successful discoverers of supernovae, such as Charlie Kowal at Palomar and Paul Wild at Berne, somehow manage to combine their supernova hunting with searches for more important things, such as comets, asteroids and possible faint satellites of Jupiter.

Supernovae are also frequently reported by amateur astronomers. They are frequently reported, but the reports almost without exception do not refer to real supernovae. In fact--and I'm being serious now--one of the bains of our existence is the myth that supernovae are there for the discovering and that just about anybody can go out and find one. If you theoreticians thought M31 were to be allocated one supernova every three or four centuries, you haven't been around here to hear the frequent phone calls from the tyros who call up in the middle of the night absolutely convinced that one of the clumps of material they didn't notice before in one of the spiral arms of M31 is: yet another supernova. In response to our grilling, they tell us how convincing their photograph of the galaxy is (it's not always M31: M51 is also a popular target), or how familiar the galaxy is to them when they point their telescopes at it. We give them the third degree. Why the devil did they not take more than one photograph? Don't they know that George Eastman puts supernovae into his emulsions just so that they show up when one is developing a pretty picture of one of the more messy Messier objects? And as for the amateur who simply likes to go out at night and look at a particular galaxy, we ask him when he last looked at that galaxy. Invariably, the galaxy is now conveniently coming up in the east in the late evening. When he last saw it, several months ago, it was setting in the west just after sunset. Doesn't he know he suffers from a slight astigmatism? The orientation of the galaxy to his eyes is now different from what it was, and that prominent little knot of stuff he sees now so clearly he never even noticed in the spring. Interrogation of this type generally disposes of all but the least trepid, although it can be extremely time consuming.

Actually, most supernovae that are discovered are relatively unobscured, and they therefore stand out as bright as the parent galaxy itself. Comparison of a new photograph (or pair of new photographs) with one taken under similar conditions in the past will usually do the trick. Photographic discoveries can often be considerably delayed, however, and for some time there has been a lot of talk (but relatively little action, it appears) by several groups interested in using fancy real-time devices for making discoveries. Justus Dunlap, at the Corralitos Observatory in New Mexico, did successfully use a television device for supernova hunting for several years, until his funding ran out.

The most productive real-time device for picking up supernovae at the moment, however, is the eyes and the memory of the minister of the Australian Uniting Church in Maclean, New South Wales. The Reverend Robert O. Evans has truly amazed us all by the way he can go out to the telescope in his backyard and know, with the speed of light, that the galaxy NGC 1532, or 1316, or 5236 or 1448 or 3169 or 1559 or 991, is right now being invaded by a supernova, perhaps one still on the rise and therefore a particularly valuable target for astrophysical observations over the whole range of the electromagnetic spectrum. Nobody else in the world can, it seems, do this: only Bob Evans, and he's already done it close to a dozen times. On a recent evening he started out his routine, just after sunset, and within a period of four hours or so he had checked out 337 galaxies against the charts carefully stored away in his head. In the 338th galaxy, NGC 7184, there was a supernova. This was of course immediately picked up by his sensors, and, as is his custom, he quickly alerted the people at the Anglo-Australian Observatory about the phenomenon. When they were reasonably satisfied that there really was a supernova, word was sent to the AAVSO and to us here in Cambridge.

Even a Bob Evans can occasionally be fooled, however, and so can the AAO astronomers in their attempts at confirmation. We were a little taken aback a couple of years ago when they came up with an eleventh-magnitude supernova in a galaxy in Aries. It was late October, and Aries was very conveniently observable in the northern hemisphere, but nobody else seemed to have noticed any supernova. The people at the AAO assured us it was there. Since we are specialists in the more important areas of astronomy I mentioned earlier, it did not take us very long to realize that the supernova was in fact asteroid number 55. Our rather facetious remark to the AAO that the object must have moved quite a bit during the four hours they were following it with the Anglo-Australian Telescope, and that its spectrum must perhaps have been somewhat devoid of emission lines, did not elicit any response, which is perhaps just as well, for the asteroid, originally discovered at the Dudley Observatory in Albany in 1858, just happened to be named Pandora.

Foreground stars can also be a tremendous source of confusion, and our office can only apologize for relaying to the astronomical community a report in January 1983 from a reputable group of professional astronomers that a foreground star in the galaxy NGC 1265 was a supernova. Then, to make matters worse, we managed to relay a report in October 1983 from a _different_ reputable group of professional astronomers that the _same_ foreground star in the _same_ galaxy NGC 1265 was: a supernova. 1265 is not a prime number, and we shall earnestly try to be wary of it in the future.

Believe me, it can be a problem to get a quick and definite assurance that a reported discovery really is of a supernova. And it can then take several weeks more for observers to decide whether the supernova is of type I or type II. We are quite frankly sick and tired of certain professional astronomers who habitually send us reports of discoveries of "possible supernovae" (as they themselves describe them) of magnitude perhaps 19 or 20, presumably on the basis of a single exposure. There never was any published spectroscopic follow-up to the supernova Bob Evans reported in July in NGC 7184. So just yesterday some Canadian astronomers were telling us that this supernova was really a foreground variable that was clearly present, though much fainter, on the Palomar Sky Survey prints. Only by publishing this negative report were we finally beginning to elicit vague suggestions that the NGC 7184 supernova is in fact real and of type I--or maybe it is of type II. Anyway, it's not an asteroid.

Coupled with the need for quick and reliable spectroscopic information concerning the type and even the existence of a supernova is the need for the accurate measurement of the object's position. The discoverer traditionally describes the position of his supernova by the offset from the nucleus of the parent galaxy. The range of offsets given by different observers for the same supernova is laughable, particularly when one realizes that many of the very same observers are also involved with the much more precise work of measuring positions of comets and asteroids. A couple of years ago, we published on one of our _Circulars_ a plea from several astronomers, concerning the need for accurate astrometry of newly reported supernovae. Such astrometry is of course essential if there is any hope that radioastronomers or astronomers working with satellites at ultraviolet or x-ray wavelengths will be able to detect the supernova. I must say that the response to the plea was quite good. I want to mention in particular the prompt accurate measurements we regularly receive from Bob Argyle at the Royal Greenwich Observatory, for I know that these are proving helpful to astronomers who have been exasperated when the discoverer has said a supernova is 20" west and 15" north of the parent galaxy's nucleus, another observer has put it 5" east and 30" south, while a third has informed us that the supernova in fact should be associated with a companion galaxy a couple of minutes of arc away.

Oh! I almost forgot those of you who have been struggling with the problem of the proving or denying the primeness of the number 4757. As I said, the clue is the number 3. If you add 3 to 4757 you get 4760, which any fool can factorize into 70 x 68. That can of course also be written as (69+1) x (69-1). It is then obvious that (69+2) x (69-2) have to be the factors of 4760-3. So the factors of 4757 are the prime numbers 71 and 67. If you don't like that approach to the problem, you can instead multiply the number 4757 by 3. This time you can use your calculators, and you ought to come up with 14 271. It is easy to see that this number is divisible by 201, and that the other factor is 71. The first factor of 4757 is therefore one third of 201, or 67.

Everyone in this room is excited by the prospect that the long-overdue supernova may soon appear in our own galaxy. If that supernova does happen to be brighter than Venus and is not too close to the sun, you will undoubtedly learn about it soon enough. But as likely as not that next supernova will be affected by obscuration and no brighter than fifth magnitude; perhaps it will be as faint as tenth magnitude. With the responses we usually get to our requests for confirmation, it may therefore well be several days or even possibly weeks before it is clear that the object really is a supernova, as opposed to an ordinary nova or even just a long-period variable.

ACKNOWLEDGMENT

The tests for the primeness or otherwise of 4757 were vividly discussed (for I remember them after all these years) by J. M. Hammersley at a meeting of the Oxford University Mathematical Society one Tuesday evening in 1957.

Vol. 195: Trends and Applications of Pure Mathematics to Mechanics. Proceedings, 1983. Edited by P. G. Ciarlet and M. Roseau. V, 422 pages. 1984.

Vol. 196: WOPPLOT 83. Parallel Processing: Logic, Organization and Technology. Proceedings, 1983. Edited by J. Becker and I. Eisele. V, 189 pages. 1984.

Vol. 197: Quarks and Nuclear Structure. Proceedings, 1983. Edited by K. Bleuler. VIII, 414 pages. 1984.

Vol. 198: Recent Progress in Many-Body Theories. Proceedings, 1983. Edited by H. Kümmel and M. L. Ristig. IX, 422 pages. 1984.

Vol. 199: Recent Developments in Nonequilibrium Thermodynamics. Proceedings, 1983. Edited by J. Casas-Vázquez, D. Jou and G. Lebon. XIII, 485 pages. 1984.

Vol. 200: H. D. Zeh, Die Physik der Zeitrichtung. V, 86 Seiten. 1984.

Vol. 201: Group Theoretical Methods in Physics. Proceedings, 1983. Edited by G. Denardo, G. Ghirardi and T. Weber. XXXVII, 518 pages. 1984.

Vol. 202: Asymptotic Behavior of Mass and Spacetime Geometry. Proceedings, 1983. Edited by F. J. Flaherty. VI, 213 pages. 1984.

Vol. 203: C. Marchioro, M. Pulvirenti, Vortex Methods in Two-Dimensional Fluid Dynamics. III, 137 pages. 1984.

Vol. 204: Y. Waseda, Novel Application of Anomalous (Resonance) X-Ray Scattering for Structural Characterization of Disordered Materials. VI, 183 pages. 1984.

Vol. 205: Solutions of Einstein's Equations: Techniques and Results. Proceedings, 1983. Edited by C. Hoenselaers and W. Dietz. VI, 439 pages. 1984.

Vol. 206: Static Critical Phenomena in Inhomogeneous Systems. Edited by A. Pękalski and J. Sznajd. Proceedings, 1984. VIII, 358 pages. 1984.

Vol. 207: S. W. Koch, Dynamics of First-Order Phase Transitions in Equilibrium and Nonequilibrium Systems. III, 148 pages. 1984.

Vol. 208: Supersymmetry and Supergravity/Nonperturbative QCD. Proceedings, 1984. Edited by P. Roy and V. Singh. V, 389 pages. 1984.

Vol. 209: Mathematical and Computational Methods in Nuclear Physics. Proceedings, 1983. Edited by J. S. Dehesa, J. M. G. Gomez and A. Polls. V, 276 pages. 1984.

Vol. 210: Cellular Structures in Instabilities. Proceedings, 1983. Edited by J. E. Wesfreid and S. Zaleski. VI, 389 pages. 1984.

Vol. 211: Resonances – Models and Phenomena. Proceedings, 1984. Edited by S. Albeverio, L. S. Ferreira and L. Streit. VI, 369 pages. 1984.

Vol. 212: Gravitation, Geometry and Relativistic Physics. Proceedings, 1984. Edited by Laboratoire "Gravitation et Cosmologie Relativistes", Université Pierre et Marie Curie et C.N.R.S., Institut Henri Poincaré, Paris. VI, 336 pages. 1984.

Vol. 213: Forward Electron Ejection in Ion Collisions. Proceedings, 1984. Edited by K. O. Groeneveld, W. Meckbach and I. A. Sellin. VII, 165 pages. 1984.

Vol. 214: H. Moraal, Classical, Discrete Spin Models. VII, 251 pages. 1984.

Vol. 215: Computing in Accelerator Design and Operation. Proceedings, 1983. Edited by W. Busse and R. Zelazny. XII, 574 pages. 1984.

Vol. 216: Applications of Field Theory to Statistical Mechanics. Proceedings, 1984. Edited by L. Garrido. VIII, 35[illegible] pages. 1985.

Vol. 217: Charge Density Waves in Solids. Proceedings, 1984. Edited by Gy. Hutiray and J. Sólyom. XIV, 541 pages. 1985.

Vol. 218: Ninth International Conference on Numerical Methods in Fluid Dynamics. Edited by Soubbaramayer and J. P. Boujot. X, 612 pages. 1985.

Vol. 219: Fusion Reactions Below the Coulomb Barrier. Proceedings, 1984. Edited by S. G. Steadman. VII, 351 pages. 1985.

Vol. 220: W. Dittrich, M. Reuter, Effective Lagrangians in Quantum Electrodynamics. V, 244 pages. 1985.

Vol. 221: Quark Matter '84. Proceedings, 1984. Edited by K. Kajantie. VI, 305 pages. 1985.

Vol. 222: A. García, P. Kielanowski, The Beta Decay of Hyperons. Edited by A. Bohm. VIII, 173 pages. 1985.

Vol. 223: H. Saller, Vereinheitlichte Feldtheorien der Elementarteilchen. IX, 157 Seiten. 1985.

Vol. 224: Supernovae as Distance Indicators. Proceedings, 1984. Edited by N. Bartel. VI, 226 pages. 1985.